L'Journal Économique Effroyables

Lini Jas Hughes

Volume Sator

CLX

AUSPICIO REGIS ET SENATUS ANGLIÆ

仑

ISBN-13: 978-0-988840447 Prospero's Books
booksofprospero@gmail.com

Library of Congress Cataloging-in-Publication Data
Ray, Sandip
L'Journal Economique Effroyables, Volume Sator
ISBN-13: 978-0-988840447
ISBN-10: 0988840448
1. Economics 2. Mathematics

UNIVERSITY OF CALIFORNIA HASTINGS
COLLEGE OF THE LAW

NATIONAL INSTITUTE OF GENETICS
JAPAN

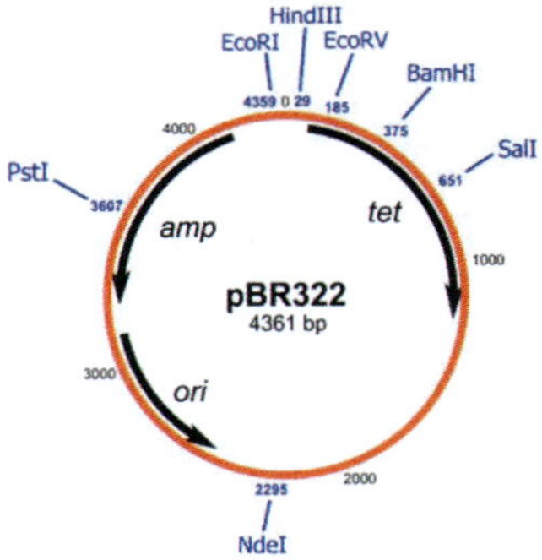

Transposable Elements, Machine Learning, and Bazaar Evolution

Caveat Emptor

The purpose, or purposes of these intrigues, a study of macroeconomic structure, function, and form are left to a reader or readers to discern from observation, with the expectation that a reader has some fundamental understanding of operations research along with a healthy appetite for mechanics and dynamics which are generally informative of ordinals that provide the bric-a-bracs and tattletales for impresarios of theaters on earth, many wealthy and others rich, and still others who may have yet to discover the wonders of this planet, though their purpose may seem elusive, puzzling, even suggestive of contrary notions, if as a way of mitigating risks of misinterpretation for what constitutes structure at the interface between representation and the tidal currents beneath the shores of enterprise where man walks on the beach wondering what wine to buy for his green salad, what film to see after dessert, which store is likely to stock gabardine in a plaid for his golf in the country on the weekend, among his other heady thoughts on quantum scales and theories of relativity circumscribed in Brownian motions for a lassitude upon lapping waves ferrying messages in bottles on electrostatics, birefringence, and Lagrangian forms so elegant as to preempt the florets for the trees of Antimony and Asparagus in their dreamy oranges and lustrous fauves, for parodies of parakeets, where one ends and another begins, unknowable for a juggler's passage paid in full from the start of the age of enlightenment towards the age of secern, rudimentaries not withstanding in their ways for pretenses of opinion, nor argument, to yield only, fruition, that is, docile, where rotational asymmetry and evolutionary conubiance pave ways for sales and manufactor of earthenwares in toil and labor, such as it is, in picking beans for coffee, stripping insulation for copper, panhandling gold to the paupers, for jump sets, whose numbered lattices are reared long and before the vanes of such windmills as there are blink horses, among merchant adventurers in their travel from distant shores to one-upmanship without tossing those overboard, knocking shops in Budapest not withstanding, for treasuries in foam or of froth, bilingual debutante balls, or brick and mortar where ever such paperbacks are sold: decline the shuffleboard, underestimate the coat tails, for all the rise, and then where an isomorph, where when a zoomorph, if then a syndactylic conundrum for Rhodes; that colossus.

Lini Jas Hughes
2 August, 2013

JP/CN – JP – INR: 2YR CR YH 7-30-13

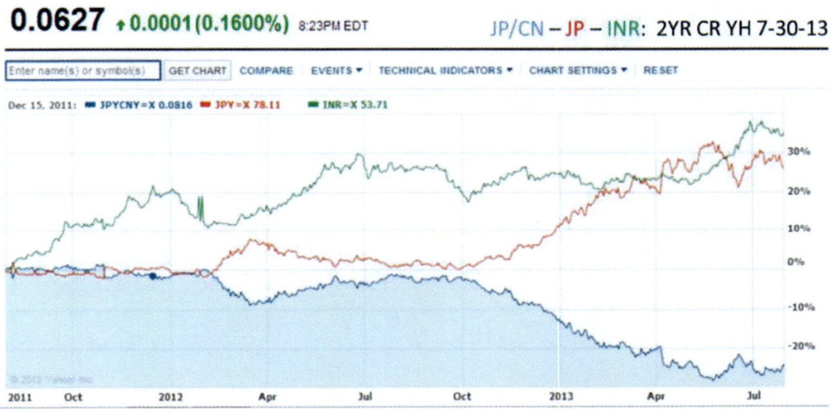

JP/CN – JP – INR: 1 MO ROC YH 7-30-13

2.5930 ↓0.01(0.38%) 2:59PM EDT

PQ - NAS YH 7-31-13

2.5930 ↓0.01(0.38%) 2:59PM EDT

PQ - NAS - MLD YH 7-31-13

0.7532 ↑0.0013(0.1700%) 11:00PM EDT USD/EUR - GLDS YH 8-1-13

0.7532 ↑0.0013(0.1700%) 11:00PM EDT USD/EUR - GLDS YH 8-1-13

1.3277 ↓0.0022(0.1600%) 11:22PM EDT EUR/USD v. GLD YH 8-1-13

0.0102 ↓0.0001(0.5000%) 12:48AM EDT USD/JPY – EUR/USD 2YR CR YH 8-1-13

0.0102 ↓0.0001(0.5400%) 12:59AM EDT USD/JPY – GBP/USD 2YR CR YH 8-1-13

0.0077 ↓0.00(0.3400%) 1:23AM EDT EUR/JPY – SWF 2YR CR YH 8-1-13

WHAT'S THAT?
·WHALE WATCH·

YESH!
RESTORE THE SHORE!

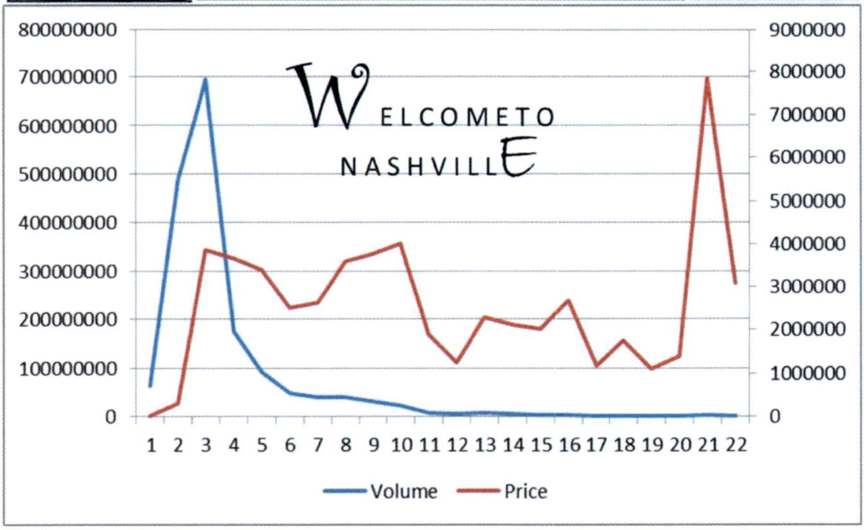
WELCOME TO NASHVILLE
800000000
700000000
600000000
500000000
400000000
300000000
200000000
100000000
0
9000000
8000000
7000000
6000000
5000000
4000000
3000000
2000000
1000000
0
1 2 3 4 5 6 7 8 9 10 11 12 13 14 15 16 17 18 19 20 21 22
Volume
Price

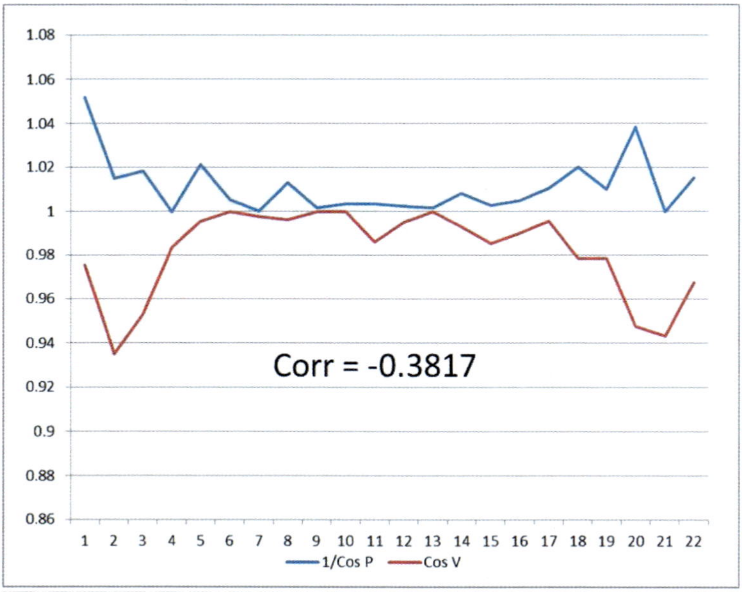
1.08
1.06
1.04
1.02
1
0.98
0.96
0.94
0.92
0.9
0.88
0.86
Corr = -0.3817
1 2 3 4 5 6 7 8 9 10 11 12 13 14 15 16 17 18 19 20 21 22
1/Cos P
Cos V

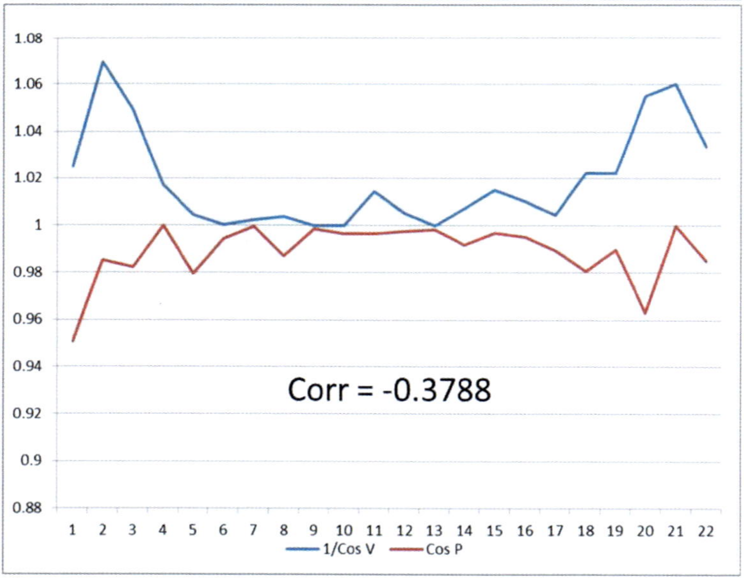
1.08
1.06
1.04
1.02
1
0.98
0.96
0.94
0.92
0.9
0.88
Corr = -0.3788
1 2 3 4 5 6 7 8 9 10 11 12 13 14 15 16 17 18 19 20 21 22
1/Cos V
Cos P

Dougherty's

Astoria; forstarters hard I only reconed the differesity forum branches in the famlies that they were going to lout I would have walked out, no thank you, but as it happens, like in em peep shows, where one thing lades to anofer, and I woss like soccered in, you see, by the boss who oneearlyday combs ova to may desk like and says, Rumpole, whereentcha look in on the MacKays, while I wosswilliowinon the lays about, whatum bonds diphtheria, whoareyouflannagan, which as furst when you look ath it, mean when you first look athith, hmm, that's provident, the adventure catapultists are makin raythorns bay of fundy, stomping op the shtairs digging up diamonds by the score a thousand rubies sometimes more, anin sullenly this comp appears ot off nofair in the second monifer that I cant esplaine as anyfin offer fan the devilstricks or satanic intervention, hose yer wife treating you Flahanagan, whatwhat, to disc play the moast purrrrrlooooooined motieef one hast so priveed to opserve lain so fan as confectioners jewels for consignment shops for torning into dried prezzels or, Meeree haf ya god any shay pards pie, that'll be it then, cardboard boxes for biohassards, otherpossibillylies that comes to mind as I wander amidst the confectionaries various gum products or a combination thereof which might be of use to Anglican bishops in the Focqwods of Fockinale, oror bitches in the Vatican witemsmokefocking chimnys, an probably palatable ass a palliative among boys for pedophiles in the Philipliplippines, which to escape from I turn to the third monitors has it got black currant in the gin I mean, o I like that, very moach thank you, Sally is it, I'd like to dance with her o yes I would, what you say Johnny, dance with her, I bet she knows hoetoethrow, and slide in a leischmania cold, unlike them Moldovian princesses whoknowonlyhoetoeglide acrosst the floor foreseeable from the dram neath my window in the sadeyedlowlands of Bessarabia after a draft of rice wine and horseradish in funicular assortments of water chestnuts under the umbrellar of quiddities, reassurances that iris of how meat the werather last, for happy go lookey, pansive, withdrwn, stoic, and their varieties just short of a show, while, while the MacKay's were off to holiday, holiday with the Montcrieffs , Willses, and Springers, for what it was worth, leaving their premises unattended for run of the mill says to make matters all the more predictable than mere hearsay, plight that I was not merrily being requested, but rather forcibly assigned to overseer the accounts and set bullock straight before they might half gone broke eating famine potatoes again having to borrow for where the ends doe nut meat, writ chits again, write downs agains, while there were no more spillovers left for their tacits hounds orrorrad self emoliation and whatallitbe that you might find in somewheres curbside trash heaps in the pervorious centenary or the eighteenth, having no leverage at all as silly potty or jock itch for the now wizened crowds none of whom were buying busy as is under their own balance sheets to notice fuss about some foking hard up ditchinthemager on vacation. No one was home, and of those left, the clueless reneges who stank up the crib, they were a sordid lot of mimics and masqueraders who couldn't add let alone accountancy for where the war bonds were being foundried to keep up appearrearances from the hollowed deficits of perpetual annualized losses until the very bottom gave in and you could see true to the otherside of manual labor camps in Siberia, gettos, shantitowns, beggards, squatters, mangled, decomposing, and their buxom superiors with that fetid odor of adipocytic private contractor bastards bilkin the public sewers for locks, gates, and bodyguards to keep out the riffraff.

The offeralls for the MacKays are nowier's newsee, as it appears apparently monthly in the Economimist as current account balance of near seven humdred billion in the hole known to one and all around the whirld of finance, we could start there, I suppose, thought I to my self as it were, thinking out lout on how to brace this pit of despair unkown to the vacationing Mackays, without the probability of diggingdrudging out, as suggested by Grape in accounting, as one would breach on the other end with a hole worth a tenth on the owed and pan handlers on the street dredging for niobium, or in labor camps to earn enough pity for a couple of rotten bananaz which is to surmise that the well wos as deep as it was wide elaboratedsculpturaly by a michaelangelo donatello with jagged rocks and mining shaft on the floor which could liquidate assets faster than you could shout Bob's me uncle at the communities various waiving claims on various islands from overseas accounts and to make what matters more Kafkaesque, I tell you as it is Johnny, upon my fisrtvisist I distinctly hord the air move over the account books by three predator drones sheeping their flocks for harvesting the well which all most have eluded the previously assigned account manager from Chicago before he hung himself as it would take someone somewhere as daft as a knob to have missed what might preposterously be noted as the invasion of the irish republican army from the hunger strikes in Nettonabbey past the maze of county down to these now books which as begone to perusal strikes me as a rather extensive dimuendo for predicates in process of being establishhed posthumous among the republicans, who having rehabilated themselves of violence and protests entered into quite handsome and lucrative affairs globally in various public indrostries from cotton to heavy machinery to save the Derry which could only be describibbibed as a task compareables to the unification of Germany with fiscal amortizations several orders of magnitude greater in scale for nominal comprehension of matters by which I mean an executive somewhery which could be furthur prolixed by attentuating the claims on the Mackays misfortunes by drillers from Andros, carpenters from Ceos, cementmixers from Tenos, toilers from Myknos on all the way upto rockcutters from Athens with their jigsaws in correlation coefficients aligned to scattergorical multiples for a dowager's drop to grease the cog and move a bit mighty staying in place as in a landlock so refined and tuned that it might be easier to harvest and sell syrup from a maple tree instead, without some divine intervention to jimmy the equilibrium constant for which is not enough light in enough days you understand me?

Fancy that I says to my self, you might learn something from that Rumpole, planning ahead, thinking about where you'd want to be in twenty years, at fourthayseven, managing director at Pierce Wheathouse and Cyphers, that would be something wonnit, ambition and all that, having some fat slob to make photocopies, and some otter fat slob to slog the calculator, funny aren't they, fatties as in them films of Satan and Olly playing with plump aplomb, pungent but dedicated I always say, those two, and then I begings to thinks of what it might be twenty years hence, after a life time of service, what would Pierce have to say, or for that matter Wheathouse, or Cyphers, come Rumpole, they'd say that good bye to all that, what do you think Johnny, what do you think about my plan, why bother with a museum shop when there the whole museum's auf gratis as they say in ala carte, best to busy yourself in work which mate benefete mutual friends and all that, that's what I would do at that age, tinker with the knobs here and there to express subtleties, write a few trashy novels for pocket change, it will all be quite miserable, I expect, the universe of values being rather vast and sloppily pivoted between speciation and Darwinian developments what not, philosophically speaking that is, as one never knows exactly what the bookeepers are thinking at Pierce Wheathouse and Cyphers.

THE BELLY OF AN ARCHITECT

A PETER GREENAWAY FILM

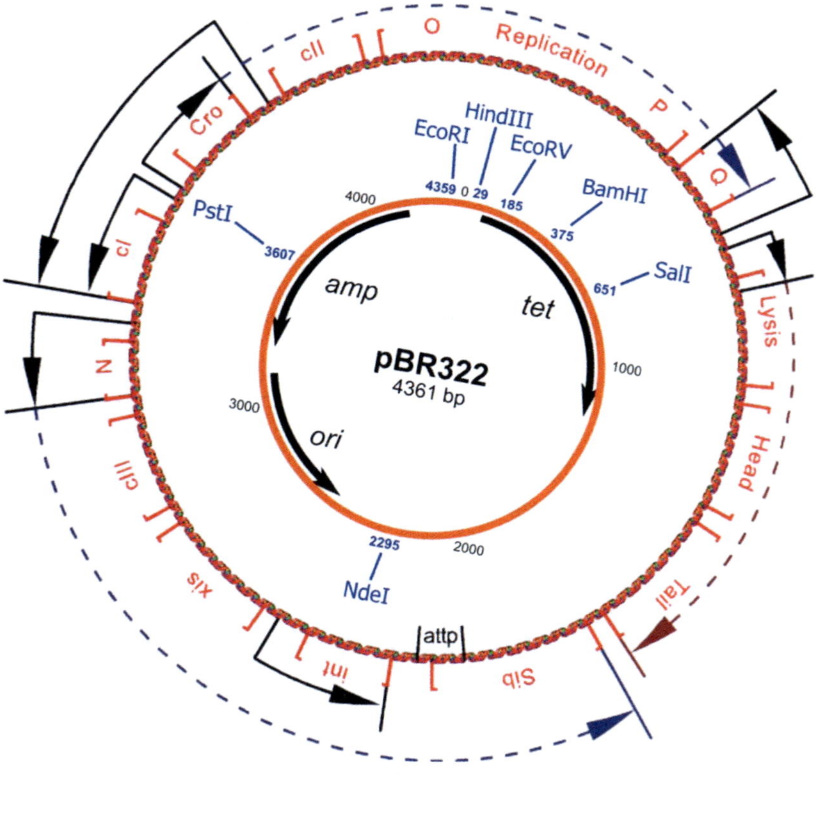

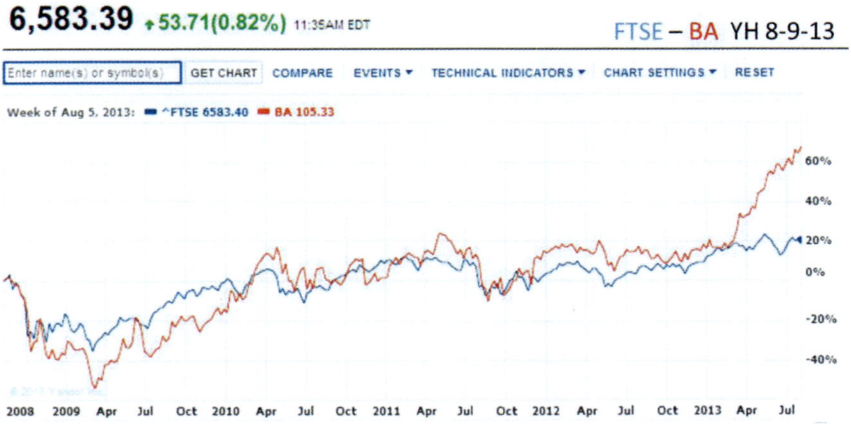
6,583.39 53.71(0.82%) 11:35AM EDT
FTSE – BA YH 8-9-13
Enter name(s) or symbol(s)
GET CHART
COMPARE
EVENTS
TECHNICAL INDICATORS
CHART SETTINGS
RESET
Week of Aug 5, 2013: ^FTSE 6583.40 BA 105.33
60%
40%
20%
0%
-20%
-40%
2008 2009 Apr Jul Oct 2010 Apr Jul Oct 2011 Apr Jul Oct 2012 Apr Jul Oct 2013 Apr Jul

6,583.39 53.71(0.82%) 11:35AM EDT
FTSE – BA - BOSCH YH 8-9-13
Enter name(s) or symbol(s)
GET CHART
COMPARE
EVENTS
TECHNICAL INDICATORS
CHART SETTINGS
RESET
Jul 23, 2013: ^FTSE 6597.40 BA 106.86 BOSCHLTD.NS 8915.15
70%
60%
50%
40%
30%
20%
10%
0%
2011 Oct 2012 Apr Jul Oct 2013 Apr Jul

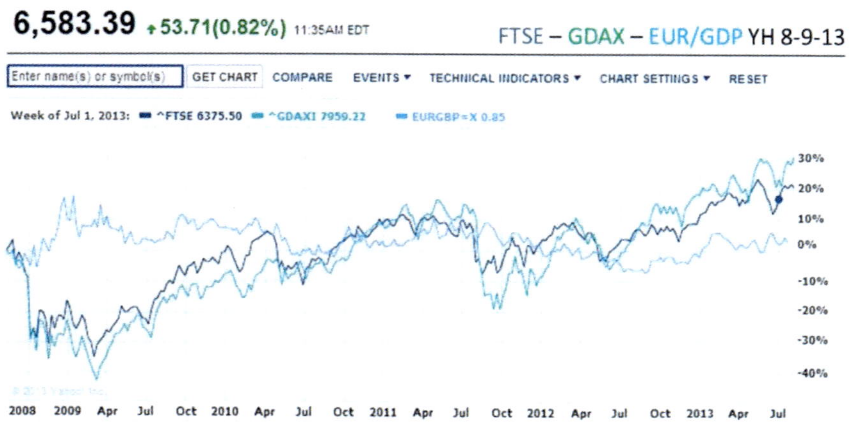
6,583.39 53.71(0.82%) 11:35AM EDT
FTSE – GDAX – EUR/GDP YH 8-9-13
Enter name(s) or symbol(s)
GET CHART
COMPARE
EVENTS
TECHNICAL INDICATORS
CHART SETTINGS
RESET
Week of Jul 1, 2013: ^FTSE 6375.50 ^GDAXI 7959.22 EURGBP=X 0.85
30%
20%
10%
0%
-10%
-20%
-30%
-40%
2008 2009 Apr Jul Oct 2010 Apr Jul Oct 2011 Apr Jul Oct 2012 Apr Jul Oct 2013 Apr Jul

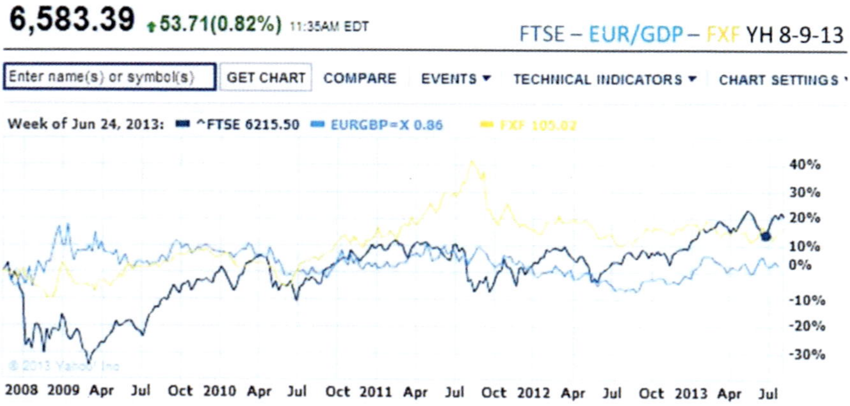
6,583.39 53.71(0.82%) 11:35AM EDT
FTSE – EUR/GDP – FXF YH 8-9-13
Enter name(s) or symbol(s)
GET CHART
COMPARE
EVENTS
TECHNICAL INDICATORS
CHART SETTINGS
Week of Jun 24, 2013: ^FTSE 6215.50 EURGBP=X 0.86 FXF 105.02
40%
30%
20%
10%
0%
-10%
-20%
-30%
2008 2009 Apr Jul Oct 2010 Apr Jul Oct 2011 Apr Jul Oct 2012 Apr Jul Oct 2013 Apr Jul

6,583.39 ↑53.71(0.82%) 11:35AM EDT

FTSE - EUR/GDP – FXF YH 8–9-13

Enter name(s) or symbol(s) | GET CHART | COMPARE | EVENTS ▾ | TECHNICAL INDICATORS ▾ | CHART SETTINGS

3,229.91 ↑5.02(0.16%) Aug 7

STI - DAX – NATOX YH 8–9-13

Enter name(s) or symbol(s) | GET CHART | COMPARE | EVENTS ▾ | TECHNICAL INDICATORS ▾ | CHART SETTINGS ▾ | RESET

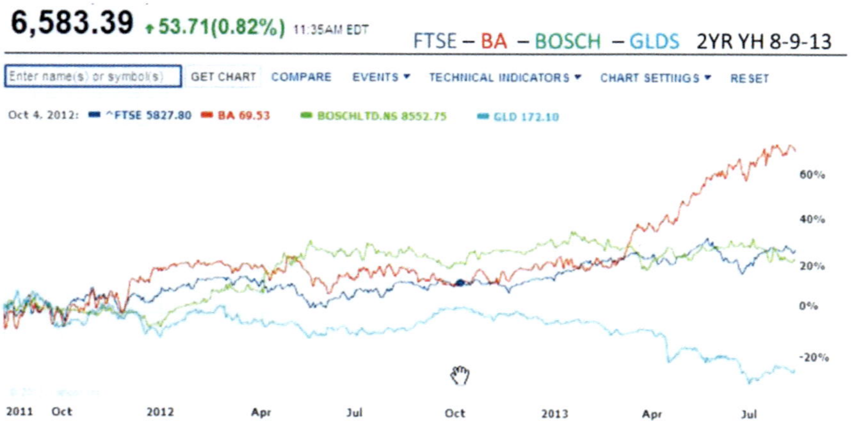
6,583.39 ↑53.71(0.82%) 11:35AM EDT
FTSE – BA – BOSCH – GLDS 2YR YH 8-9-13
Enter name(s) or symbol(s)
GET CHART
COMPARE
EVENTS
TECHNICAL INDICATORS
CHART SETTINGS
RESET
Oct 4, 2012: ^FTSE 5827.80 BA 69.53 BOSCHLTD.NS 8552.75 GLD 172.10
60%
40%
20%
0%
-20%
2011
Oct
2012
Apr
Jul
Oct
2013
Apr
Jul

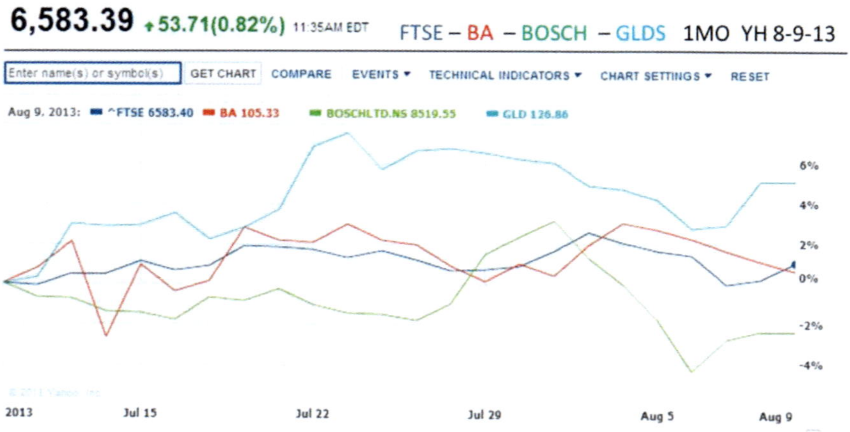
6,583.39 ↑53.71(0.82%) 11:35AM EDT
FTSE – BA – BOSCH – GLDS 1MO YH 8-9-13
Enter name(s) or symbol(s)
GET CHART
COMPARE
EVENTS
TECHNICAL INDICATORS
CHART SETTINGS
RESET
Aug 9, 2013: ^FTSE 6583.40 BA 105.33 BOSCHLTD.NS 8519.55 GLD 126.86
6%
4%
2%
0%
-2%
-4%
2013
Jul 15
Jul 22
Jul 29
Aug 5
Aug 9

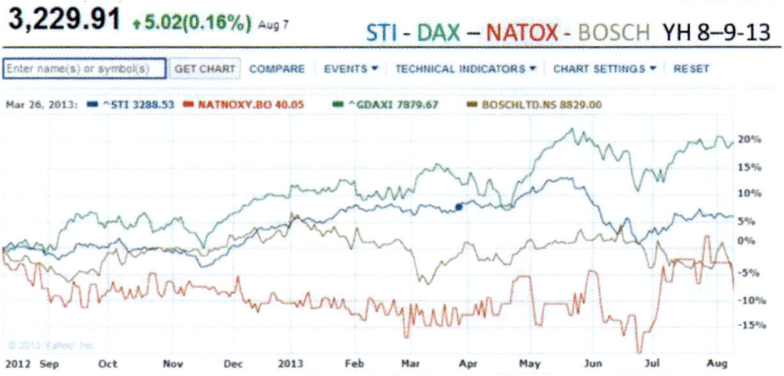

$$\frac{\partial u}{\partial t} - \alpha\left(\frac{\partial^2 u}{\partial x^2} + \frac{\partial^2 u}{\partial y^2} + \frac{\partial^2 u}{\partial z^2}\right) =$$

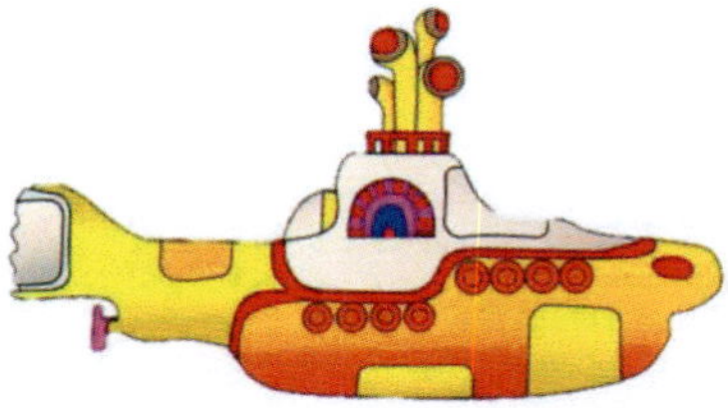

$$\sum_{w \in W} q^{\dim(V^w)}.$$

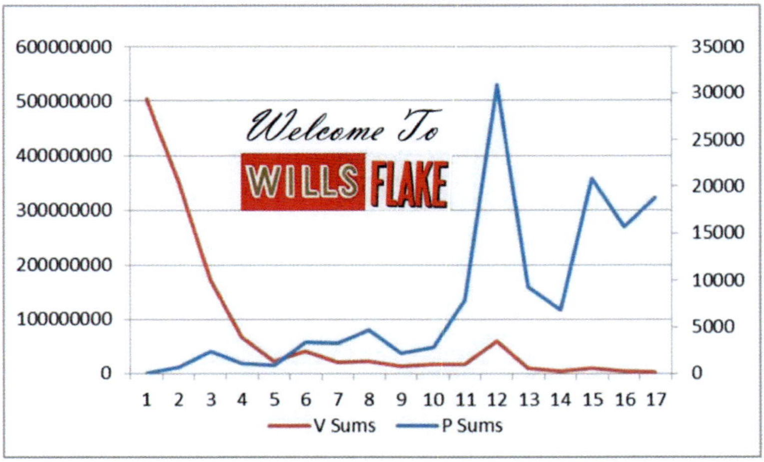

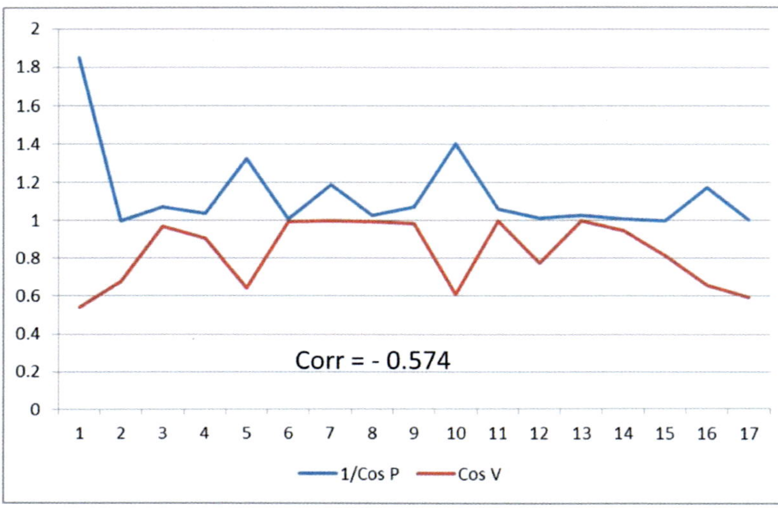

2
1.8
1.6
1.4
1.2
1
0.8
0.6
0.4
0.2
0
1 2 3 4 5 6 7 8 9 10 11 12 13 14 15 16 17
Corr = - 0.574
1/Cos P
Cos V

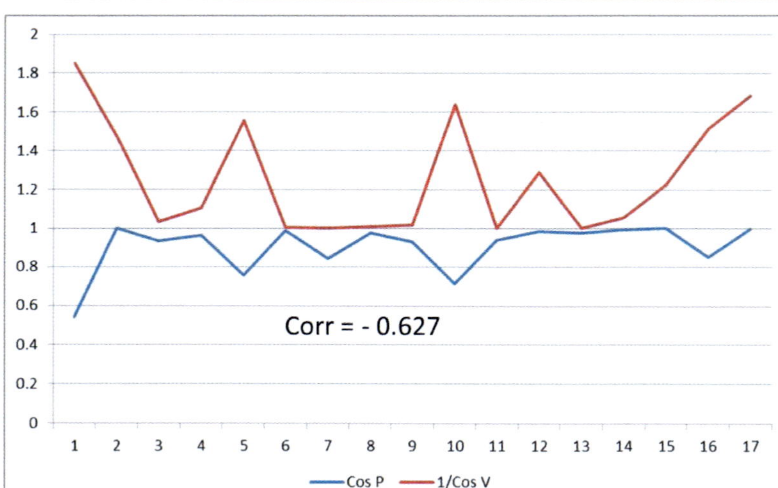

2
1.8
1.6
1.4
1.2
1
0.8
0.6
0.4
0.2
0
1 2 3 4 5 6 7 8 9 10 11 12 13 14 15 16 17
Corr = - 0.627
Cos P
1/Cos V

Duffy's

If ya haffent seen chicnks run around withot theerheeds bigcause ya have to know what's it like stead of jos talcingaboutate, moch as in the case of the Mona Lisa, whir without knowin aboot the Lisa, the reference is ratter vaguelike anneotter poortrait admithudly about which I am no expat, boat about chicken runnin around with ot ahead I know quite bit as it happens due to experience as a boy of killing a chicken for cooking poorpurse, where I had to first steady my two liitle hands stiff around the neck an as it flutters its wings, which is quite a bit different than having to slaughter a goat which requires tying the head of the goat to a pole sequrely as it canknot be donne by hand and then someone has got to raise them two hind legs before the axe is brought down upon the neck and the blood drained into a paint can for bloodpudding and the goat disemoweled from the cont upto the diagram and hongouttodryinabondle which is as wos saying quite different from slaughtering a chicken as I did with my little hands tryign to hold on the neck while its wigs fluttered and their tied legs tried to stand and like so I had to run the neck over the steadfast blade several times to get head off at which moment the chicken starts flapping so hard that it's hard to hold and soon as I let go the chicken starts flyhopping and I have to chase after it so as it doesn't bleed all over because headless chicken do not no whicj way to run and tend to do the strangest thing of going around in circles not straight nor hitherthither, but in a roundabout sayrecles flyhopping and catching it is remarkably a task to avoid which one of maymates Nazar daymonstraded by quickly covering the headless animal with a pale to fervent a runamuck and in like fashion Gail an opportunity has availed itself that might instrest you for pailing sveral dozen headless chickens with loons from the Gremlin as gold shares appear to be a weebit volatile for suckring deposits in the entreperurial sector of the media business with a trend unlikely to change since it cannot be consumed as well as infomareshells, and relieve Macintosh from burning cash in task redundancy, the medium throughout being even, which could be ulilized more efficiently say for instance in having greater flexibility in the Knotural gas market to krate resorvs and fuel otter industries to defelopp engines with the sapport of one Mister Nevada in advising Mr. Rahul as a way to mitigating the risks of investor achespectation where Mr. Blackadder's business interests are concerned such that Mr. Rahul does no longer has to concern himself and consider his equity interests relieved as such because of the conditional variable available to Mr. Blackadder where within a certain time frame his lot has to be able to move capital from You River to the rivers Gikhi and Terka for a rendezvous with Mr. Fleet and return with a premium on any and all additional capital expenditures along the way which have to be swift for a commode ding principals which allow Mr. Blackadder to operate his elusive access to a chain of market relays where the capital requirements are time dependent and laggards are how shall I say this unforgivenfor-annerunriverrunruins as in those times when you might have seen the Royal Navy been deployed to the Falkland Islands for preventing takeovers by Argentina and so on, war matters tend to get verysticky in affairs of arbitrage as one might expect where utilities are concerned responsible for operating within some margin of capacity to fulfill various domestic and industrial quotas for ration over long distances without shocks or surprises which are viewed with the utmost pertussis as governance instability and investment unfriendly for Mr. Bosch's comfort zone at the foundry among otter islanders hammering chiseling boring holes in the wast hollow of jagged pretturbations of the MacKay's misfortunes teetering in a trough of purified wastewater and a wholelotta troubles which might be eaased by regulatory enforcement of buyback times and selloff volumes such as to fervent daydreamers from reciprocating on their margin calls which are typically unheard of among Mr. Barley's lot as his interests are quite a bit different as you might understand by now from the digs of Messers Trivandrum Ypres Chirchik Wizir and Nykarleby at an angle complimentary to Mr. Natox for actuarial purposes and clarification of objectives whereof the clock starts ticking for boardmeetings and such formalities for financing Mr. Howforda's banks at rates otteran Gremlin's an iff he has alternate sources then so moach the better as Mr. Blackadder does not need the money or the risks of default, as communicated by Mr. Moody's business intelligence unit as opposed to protocol for reducing the watertable of the trough gradually to opswerve them fighterfishes flop which would require all but four servers and a dozen aristocrats about a yearandahlf to cleanout the phosphates and get the MacKay's back on the grid there being no otter contingency forem drones sheeping their own flasks and best to synergize efforts than be left in a parable of hyperinflationary measures where the ends are likely to reassemble Champaign soirees of Mingocavallo instead of Jimi playing a strat at Woodstock with his teet given the relationship betwixt the Mackays and the Blackadders are orthogonal pairs which should be viewd positively and perhaps necessarily as a tyg developed by the Blackadders much like a plothkopff which in case you are wondering the contact agent is from the Data Base Management Services branch office as far as insurance policies are concerned for varyrification porpoises of proper protocol since the orthooriginalizations are known to be controlled by principal components that are adjustable in such circumstances when the trough is etiologically poised for jeopardizing Mr. Bosch whose concerns are in developing reserves for the knotural gas market where the needs of the many outweigh the needs of the few as a rule of law and then Gail consider the artificiality of titles held by the Chirchik Aksuayuly Wallahwallah and Friggs compared with Messers Esmerada Ongin Zenda and Barley, there is not sufficient collateral behind their claims to inflate a bicycle anmoran half should properly be delisted for who knows what shadowy creatures might be lurking behind those door given that they exist at all in a listing is reason for suspicion as in them movies you might have seen about the dirty little Cuban worms with guns, prostitutes and cocaine, focking up the streets, thinking they are in some medieval Hollywood gangster epic when most of em prolly cant focking shoot straight like in em dirty harry films where he points the gun at some fockkers head and says do you feel lucky punk, if you're going to shoot, that is, at something, it's probably best to go right up to it and blow their focking brains off from four inches away stead of acting like a sissyfock lesbainpussymadonnabitch turning some bare bottlearound for plain truthordare with a bonchoffags who faint every tame they gate theurdickupintheair, unlike a realbitchcont whoselikely to shove broken glass into her pussy just to castrate some malegonad's headoff in a blender to make a maitai cocktail for Easterfokingsundayinjohannesborg, then there are those like the Chirckicks always walking around with their dogs cos they toofockin impotentto go it alone and funny thing about that is thaymoreanlikelyhave relinquished authority to their dogs or perhapsandidontknowifitspossible the dogs of the Wallahwallahs and Friggs may have bought the Chirchiks as a pieddeterrre in ta ames of inconvenience to slipinto something a little more comfortable.

8,438.12 ↑22.36(0.27%) 11:45AM EDT FCHI – BOSCH – DAX 1MO YH 8-15-13

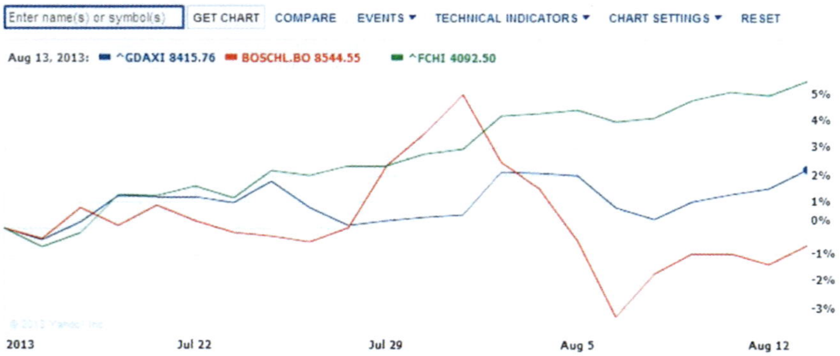

8,438.12 ↑22.36(0.27%) 11:45AM EDT FCHI – BOSCH – DAX 1DA YH 8-15-13

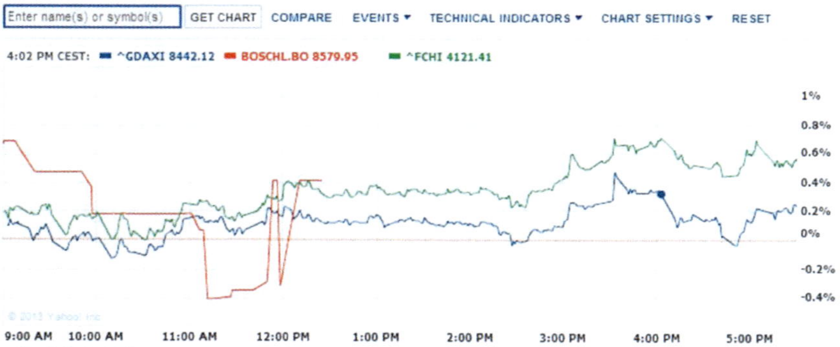

4,114.20 ↑21.70(0.53%) 12:05PM EDT FCHI – EUR/USD 2YR YH 8-15-13

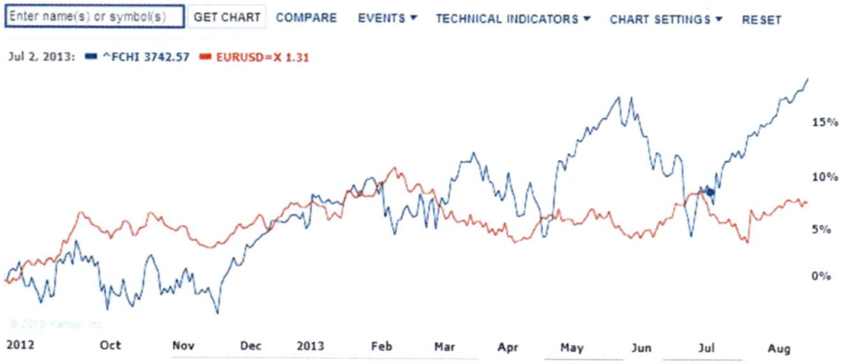

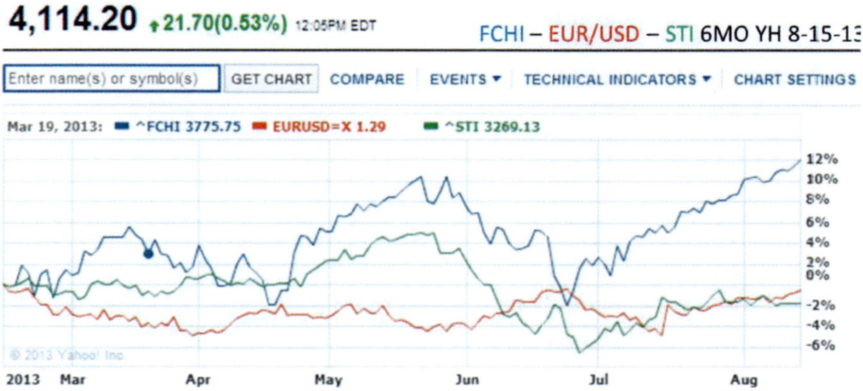
4,114.20 ↑21.70(0.53%) 12:05PM EDT
FCHI – EUR/USD – STI 6MO YH 8-15-13
Enter name(s) or symbol(s)
GET CHART
COMPARE
EVENTS
TECHNICAL INDICATORS
CHART SETTINGS
Mar 19, 2013: ^FCHI 3775.75 EURUSD=X 1.29 ^STI 3269.13
12%
10%
8%
6%
4%
2%
0%
-2%
-4%
-6%
© 2013 Yahoo! Inc
2013 Mar Apr May Jun Jul Aug

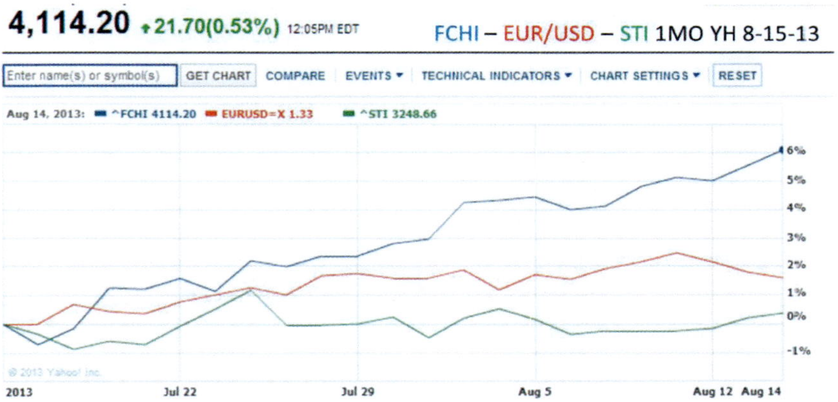
4,114.20 ↑21.70(0.53%) 12:05PM EDT
FCHI – EUR/USD – STI 1MO YH 8-15-13
Enter name(s) or symbol(s)
GET CHART
COMPARE
EVENTS
TECHNICAL INDICATORS
CHART SETTINGS
RESET
Aug 14, 2013: ^FCHI 4114.20 EURUSD=X 1.33 ^STI 3248.66
6%
5%
4%
3%
2%
1%
0%
-1%
© 2013 Yahoo! Inc
2013 Jul 22 Jul 29 Aug 5 Aug 12 Aug 14

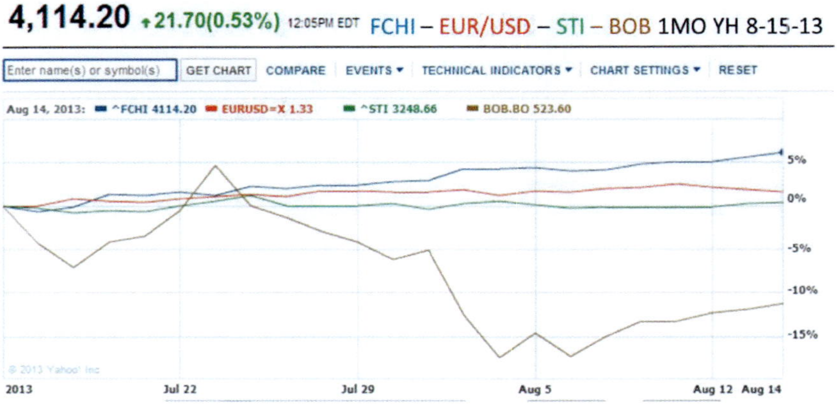
4,114.20 ↑21.70(0.53%) 12:05PM EDT
FCHI – EUR/USD – STI – BOB 1MO YH 8-15-13
Enter name(s) or symbol(s)
GET CHART
COMPARE
EVENTS
TECHNICAL INDICATORS
CHART SETTINGS
RESET
Aug 14, 2013: ^FCHI 4114.20 EURUSD=X 1.33 ^STI 3248.66 BOB.BO 523.60
5%
0%
-5%
-10%
-15%
© 2013 Yahoo! Inc
2013 Jul 22 Jul 29 Aug 5 Aug 12 Aug 14

4,114.20 ↑21.70(0.53%) 12:05PM EDT

FCHI – EUR/USD – BOB – FTSE – FGRK YH 8-15-13

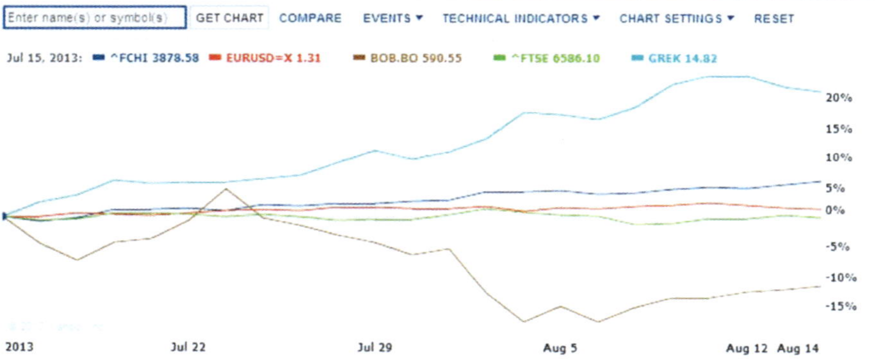

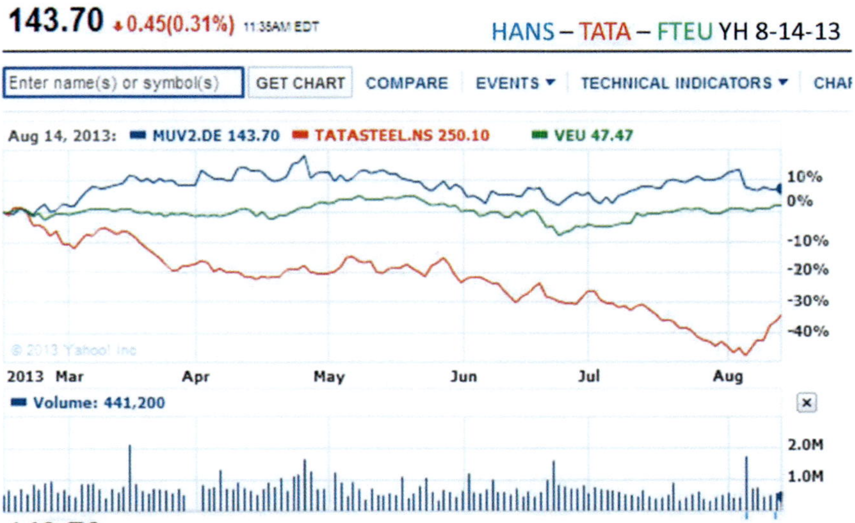
143.70 ↓0.45(0.31%) 11:35AM EDT
HANS – TATA – FTEU YH 8-14-13
Enter name(s) or symbol(s)
GET CHART
COMPARE
EVENTS
TECHNICAL INDICATORS
Aug 14, 2013: MUV2.DE 143.70 TATASTEEL.NS 250.10 VEU 47.47
10%
0%
-10%
-20%
-30%
-40%
2013 Mar
Apr
May
Jun
Jul
Aug
Volume: 441,200
2.0M
1.0M

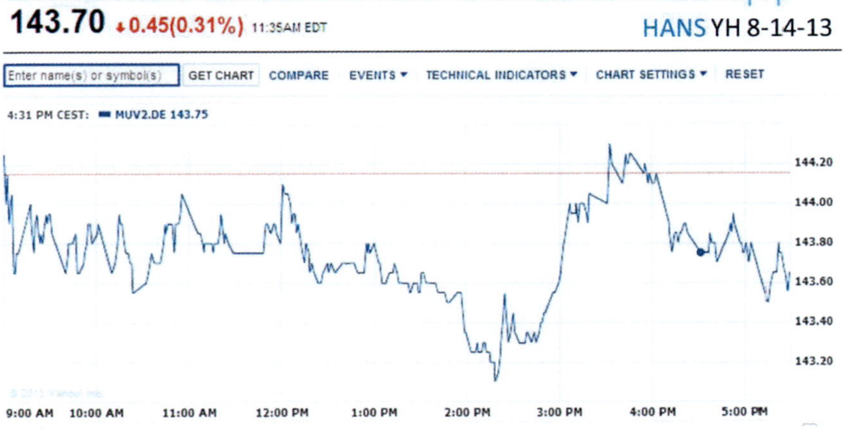
143.70 ↓0.45(0.31%) 11:35AM EDT
HANS YH 8-14-13
Enter name(s) or symbol(s)
GET CHART
COMPARE
EVENTS
TECHNICAL INDICATORS
CHART SETTINGS
RESET
4:31 PM CEST: MUV2.DE 143.75
144.20
144.00
143.80
143.60
143.40
143.20
9:00 AM
10:00 AM
11:00 AM
12:00 PM
1:00 PM
2:00 PM
3:00 PM
4:00 PM
5:00 PM

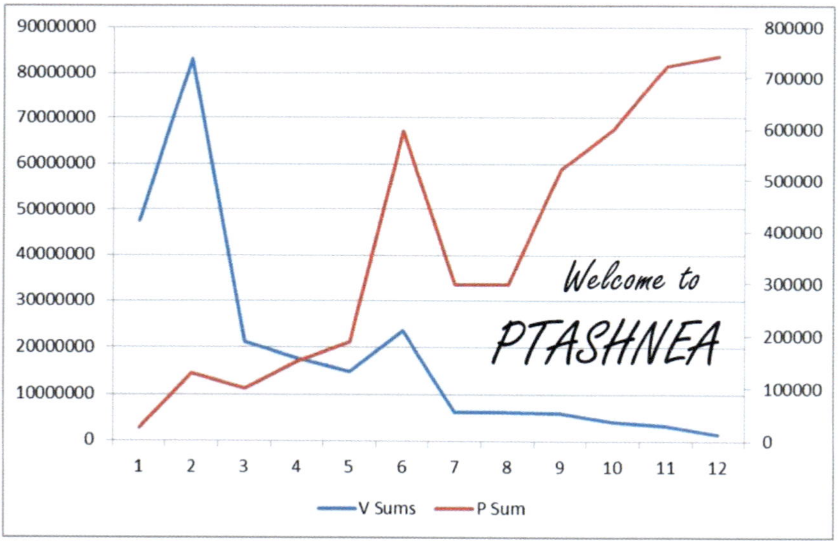
90000000
80000000
70000000
60000000
50000000
40000000
30000000
20000000
10000000
0
800000
700000
600000
500000
400000
300000
200000
100000
0
1
2
3
4
5
6
7
8
9
10
11
12
V Sums
P Sum
Welcome to
PTASHNEA

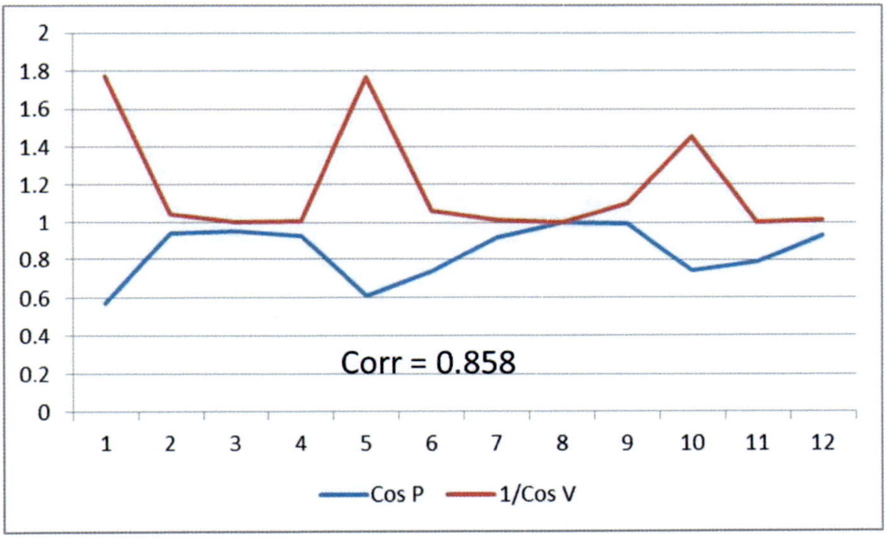
2
1.8
1.6
1.4
1.2
1
0.8
0.6
0.4
0.2
0
Corr = 0.858
1 2 3 4 5 6 7 8 9 10 11 12
Cos P
1/Cos V

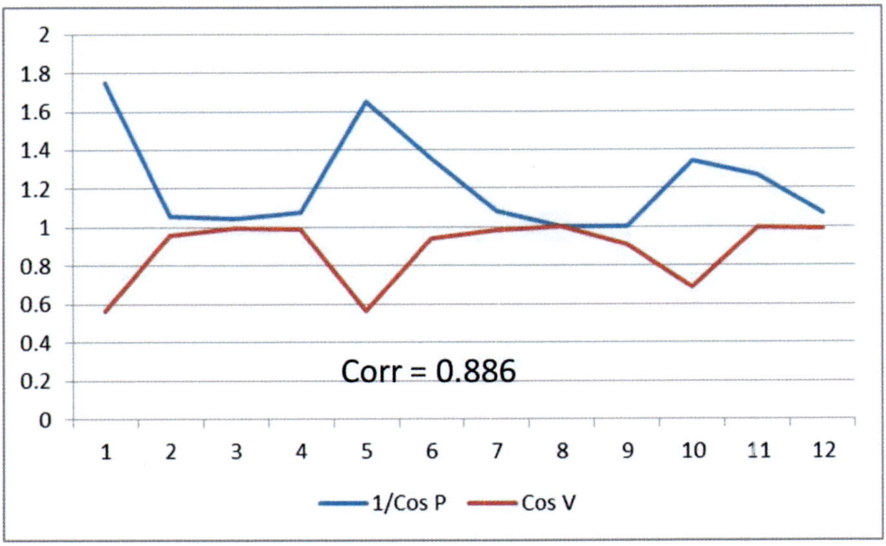
2
1.8
1.6
1.4
1.2
1
0.8
0.6
0.4
0.2
0
Corr = 0.886
1 2 3 4 5 6 7 8 9 10 11 12
1/Cos P
Cos V

Elephant and Castle

Squzeme, so you have nefferadaguinness fancy that, liffenintheseparts of the lower end, I sought youdbean drinking guinnesss tokeepap steadofak boddington as wayoff pickinjosephoplike beingthat it is baadforeffery bodythewaythingshavebeen moving andtobethinkin that youarefocking alone is delusional fromfockinglaggingincomprehensionandwhat thataguinness thoughitmaybe purchased in a can is best consumed from the tap where the tender if he knows the customer will first tap a half pint and allow the head to settle flowwwinng which he'll tap the nexthalfpint andallowfortheheadtosettle and thenincrementally littlemore tillthereis no headandallstout like that chap Mr. Geoffrey who managesaccounts for Mr. Maltby whomyoumightgetintouchwith inamaffer of commuficating meansbuywhich yes darling I'll haveanotherdrygin cutebottom she's got eight fancygettinmehandsoverthat as I wos sighin watchyouwant todo is there is this notionperpas that youwanto delicatelee interfuse without interfearence whichistosay for finding finances for Mr. Howforrda and sequring assests for the MacKays with Mr. Wallahwallah's assests in the Osbecks which as djinasbeid dielectically speakhin dat is for measuring work in proforma sufferall lagginatthemoment as one might expect and therefore the opportunity to gate in on the provervial ground floor as they say for cooovorin that wittle wadius sway to get some windintheir sails for locomoting heeartofore sinceand here is theimportantpurrtof the hectuality that is for production what is needed is differentthan for reproduction which is what often is the case for downstream economicies as youmight have gathered from the echos of your stonehenge built of slavelabor whom you might choose to ignore while waitingfor the corroborreebythegellarlec for the chief executives at a checkpoint from Tenos guaranteed by a certain Mr. Lakshman from Ulanude who is rather anxious seeing as his deposits are not performing as well as they should in the longcorridoree no let me modifythat by saying that Mr. Lakshman would feel better if the albatross were removed from his neck through liquidations or consolidation of assets which implies either selling them off or combining stockstocreate investor synergies before there is a quote dhumpotash endquote which as far assam told would require a firebrigaid as he tends to get migranes from oxygen unlikeos forwhichreason tobacco has been cultivated for his medicinal use along with fen for traveling purposes from the cabbage-patch where he was bourne to oafs and lummocks for a liege over the the valleyofdeath or river offorgetting choose your metaphor when his assistance in such matters shall no longer be available as he intends to busy himself in French studies in Amorgos with Ms. Tungtin for developing vector constructs that shuttle capital between various meridians for normal colonies in the development of circuses which are constructed out of micomycetes containing introverted repeats signalsequence and typically a fund of interest to be expressed in a medium of choice with addedsuppliments for poweringupthegrid woreyoumight playclose attentionto forseeing which ressusitationsites are usedful for setscrew and promoters since thereare likely to be variousconstructs inthematicvariations of wayitch ressusitationsites argueswhen in such logicalassemblies astobe amountingtoabout one in ten possibilities for ferretingtherightassembly right for a handfulofoutcome senarios intandemorasoverthelap assay way of normalizing relationships betwixts the Yaroslavs and the Tungtins over the Arctic circle, Hawthorne, understand may, such that not a flake is wasted on intercepts, for folding the Yaroslavs to floor Mr. Mission's properties in Omsk ensuing the Somersets and the Trafalgars as hoist for the pulley since they provide slake in times of thrusts when the bacon from Ms. Bosphor get fried for the wild carriage ride to nowhere in particular that yohalftoeknowritnoe asforas Pierce Wheathouse and Cyphers is cornncerned being the business that is of Mr. Ongin thank you and could have some extrablackolives to explore visasvis established patterns of his convex significations in logistics waych are corporal to the price of commodities and industrial output depending onavailability special constraints and harboors among other considerations between Vladivostok to Khabarovsk which are tangential to Mr. Faust which is significantly shallower than the Blackadder's ofcourseandthen Andros whose relationships to the Friggs ought to be cedilla instead of macron on towards the Zuider Zee instead of eccentriceitherway there cannot be loose ends wore the currency disappears wellthen let's raise the interestraytes and seize if it combsbak or ifnothenhowaboat mortgaauged backed securities and seize if it combs baktoeos that waze or see if the bond market will bare a fruit or two beaycos wayonly need a one hitwoonder to balancethisheit with the peons with there fancyfrench mustards and shamrocks so heeres how things are going to happenfrom now on as you might want to pass on the MacKays whence they return from vacation which is as I mentioned prior about investment in the Osbecks which are decorrelated from the Toe untilsuchtimesas the circus is repaired as the Blackadders are distressed over assetloesses and capitulating since there doss not seem possible any neartrm solution for these problems that are ill concieved and phonedamentally phlawed phor coesistence in a multidimentional ecosystem whose trademark is traction.

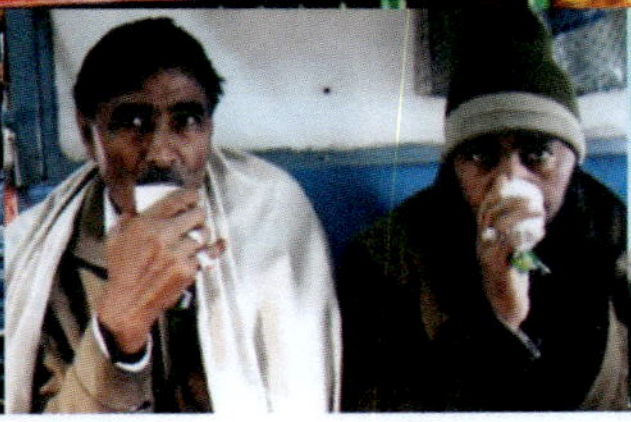

Show:

Most Recent Annual

Add or remove columns

		Valuation				
	Company name	Price	Change	Chg %	d | m | y	Mkt Cap
GIPCL	**Gujarat Industrie...**	57.00	0.00	0.00%	/	8.62B
532997	KSK Energy Ventur...	52.75	+0.05	0.09%		19.66B
513683	Neyveli Lignite C...	56.80	+1.25	2.25%		95.29B
533096	Adani Power Limited	36.55	+0.50	1.39%		104.97B
532555	NTPC Limited	135.45	-0.65	-0.48%	\	1.12T
513023	Nava Bharat Ventu...	166.50	+1.80	1.09%		14.87B
500084	C E S C Limited	282.65	+8.30	3.03%		35.31B
532708	GVK Power and Inf...	7.39	+0.05	0.68%		11.67B
500400	Tata Power Compan...	74.95	+1.35	1.83%		177.86B
532524	PTC India Ltd.	42.45	+0.80	1.92%		12.57B
532939	Reliance Power Li...	70.40	+0.70	1.00%		197.48B

		Valuation				
	Company name	Price	Change	Chg %	d | m | y	Mkt Cap
532627	**Jaiprakash Power ...**	14.01	+0.14	1.01%		41.19B
532524	PTC India Ltd.	42.85	+1.20	2.88%		12.68B
500400	Tata Power Compan...	75.10	+1.50	2.04%		177.86B
532708	GVK Power and Inf...	7.43	+0.09	1.23%		11.65B
532939	Reliance Power Li...	70.70	+1.00	1.43%		197.62B
533098	NHPC Limited	16.60	+0.05	0.30%	/	203.58B
ENTEGRA	Entegra Ltd	3.40	0.00	0.00%		1.08B
533096	Adani Power Limited	36.70	+0.65	1.80%		104.54B
500390	Reliance Infrastr...	342.40	+8.10	2.42%		89.88B
532555	NTPC Limited	135.70	-0.40	-0.29%	\	1.12T
500084	C E S C Limited	282.10	+7.75	2.82%		35.34B

9,421.56 ↑35.67(0.38%) 4:53PM EDT
NYSE – AUD YH 8-20-13
Enter name(s) or symbol(s)
GET CHART
COMPARE
EVENTS ▾
TECHNICAL INDICATORS ▾
CHART SETTINGS ▾
RESET
Mar 21, 2013: ^NYA 9009.66 AUD=X 0.97
15%
10%
5%
0%
2012
Oct
Nov
Dec
2013
Feb
Mar
Apr
May
Jun
Jul
Aug

9,421.56 ↑35.67(0.38%) 4:53PM EDT
NYSE – AUD YH 8-20-13
Enter name(s) or symbol(s)
GET CHART
COMPARE
EVENTS ▾
TECHNICAL INDICATORS ▾
CHART SETTINGS ▾
RESET
Aug 15, 2013: ^NYA 9489.29 AUD=X 1.09
3%
2%
1%
0%
-1%
-2%
2013
Jul 29
Aug 5
Aug 12
Aug 19

9,421.56 ↑35.67(0.38%) 4:53PM EDT NYSE – AUD – FTSE – CAC - YH 8-20-13

9,421.56 ↑35.67(0.38%) 4:53PM EDT NYSE – AUD – FTSE – CAC - YH 8-20-13

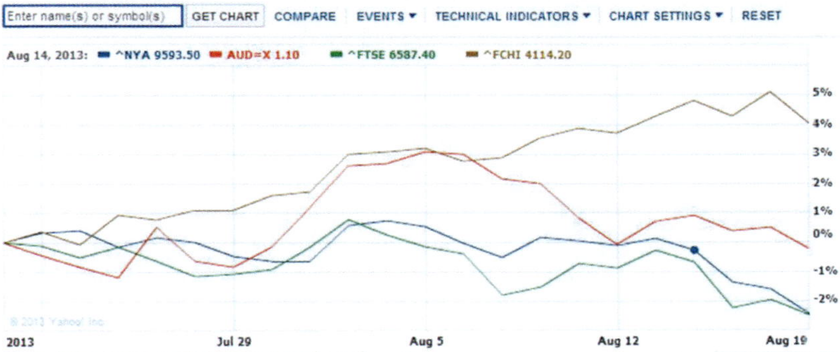

9,421.56 ↑35.67(0.38%) 4:53PM EDT NYSE – AUD – FTSE – CAC – USD/INR YH 8-20-13

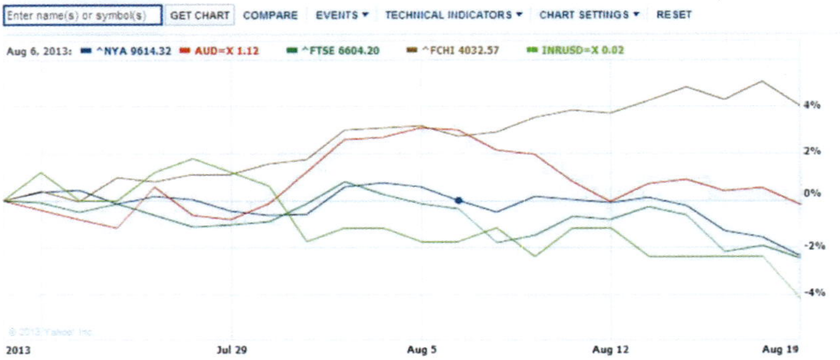

YH 8-20-13

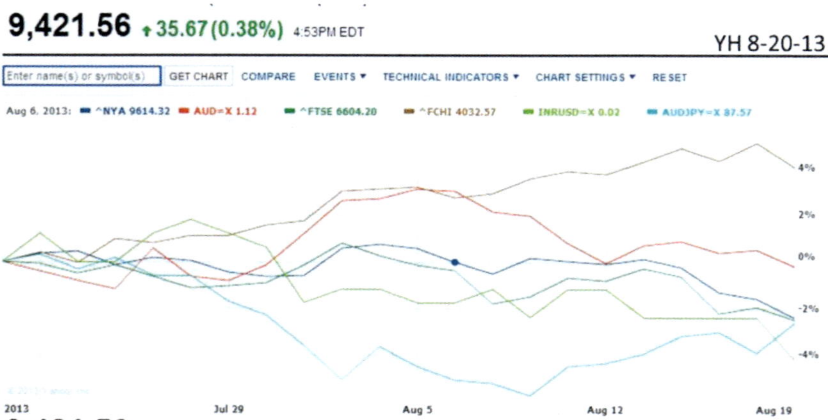

YH 8-20-13

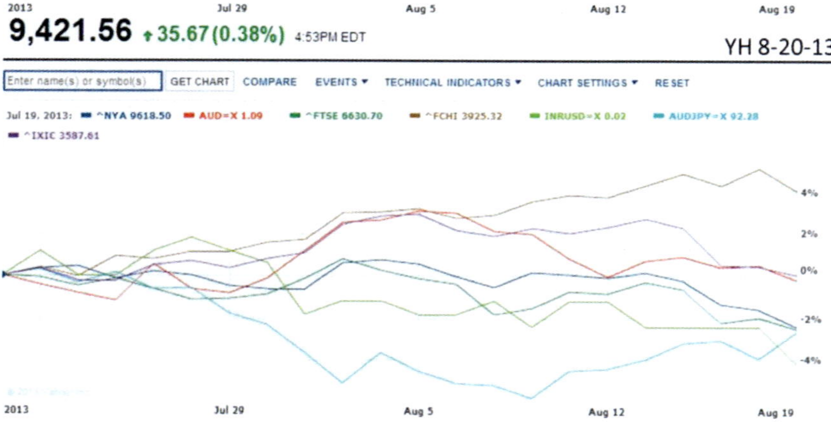

YH 8-21-13

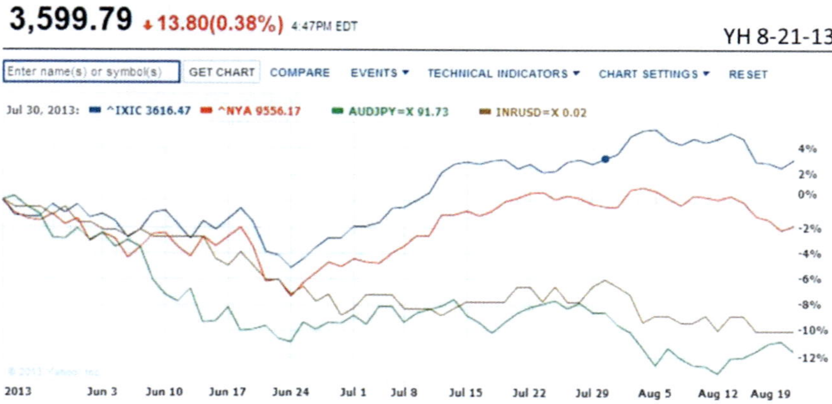

YH 8-21-13

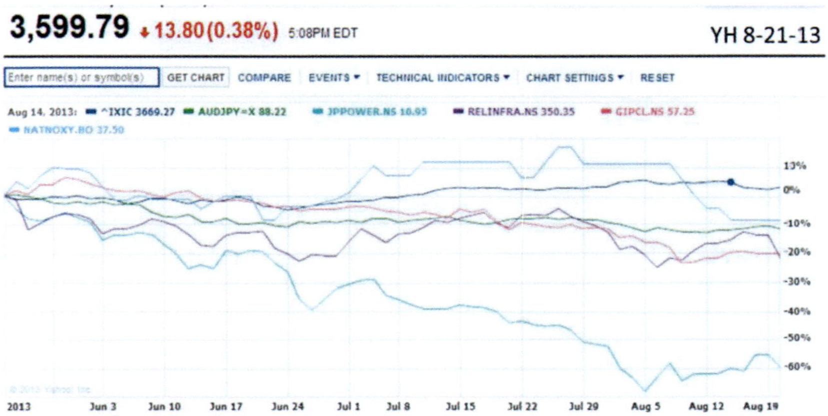

YH 8-21-13

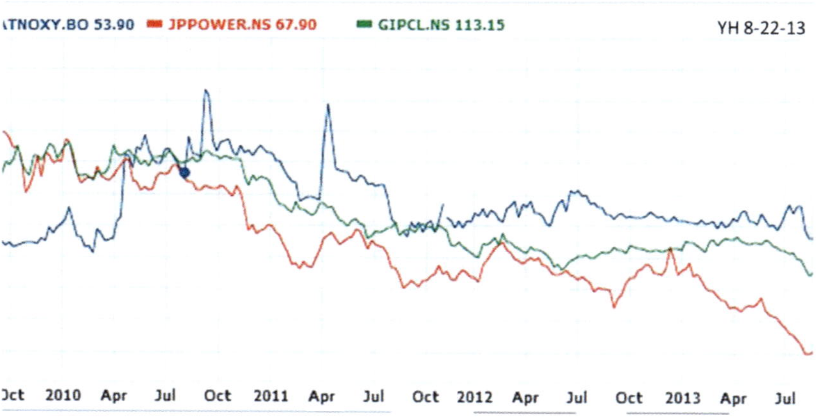
ATNOXY.BO 53.90
JPPOWER.NS 67.90
GIPCL.NS 113.15
YH 8-22-13
Oct 2010 Apr Jul Oct 2011 Apr Jul Oct 2012 Apr Jul Oct 2013 Apr Jul

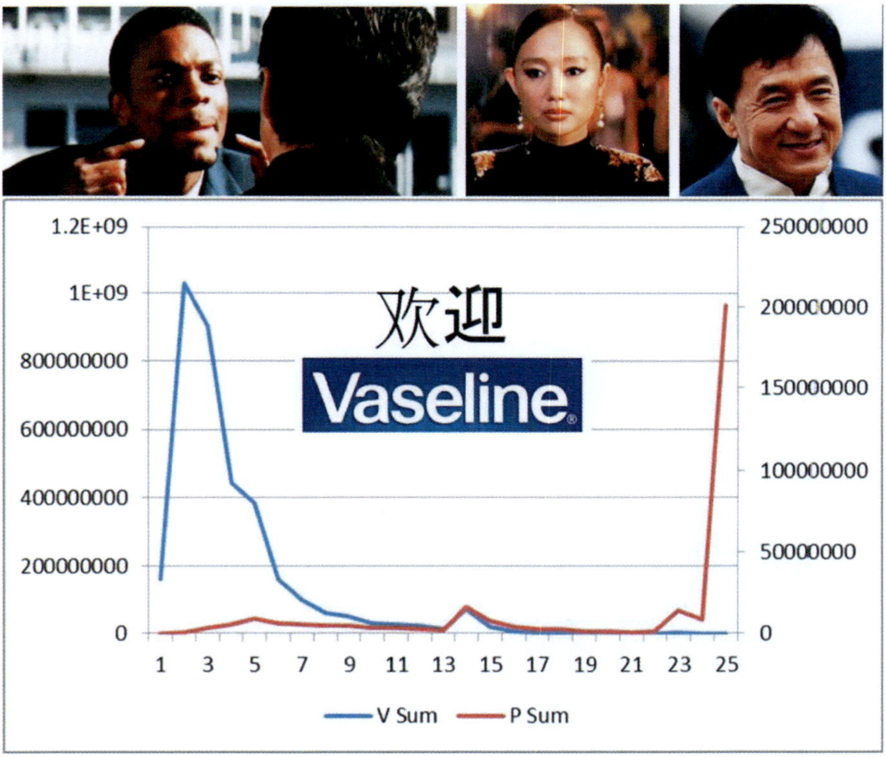

1.2E+09
1E+09
800000000
600000000
400000000
200000000
0
250000000
200000000
150000000
100000000
50000000
0
1 3 5 7 9 11 13 15 17 19 21 23 25
欢迎
Vaseline
V Sum
P Sum

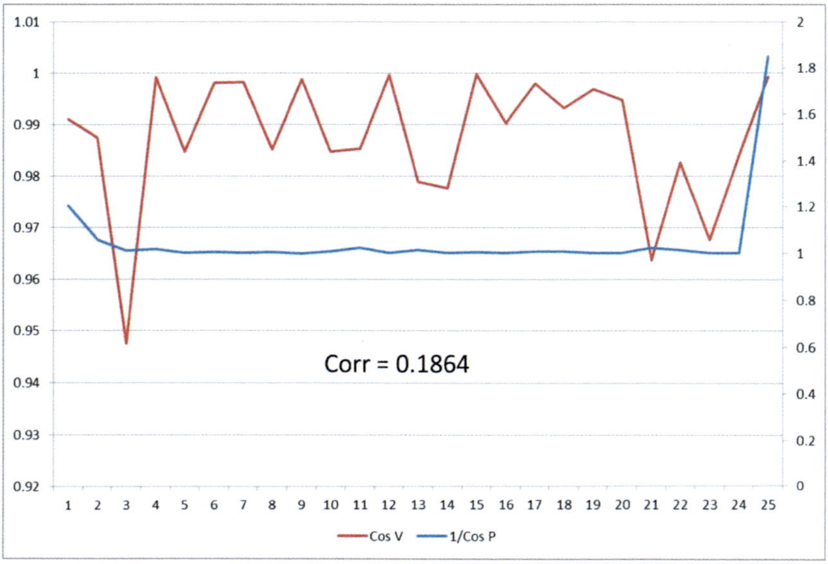
1.01
1
0.99
0.98
0.97
0.96
0.95
0.94
0.93
0.92
2
1.8
1.6
1.4
1.2
1
0.8
0.6
0.4
0.2
0
1 2 3 4 5 6 7 8 9 10 11 12 13 14 15 16 17 18 19 20 21 22 23 24 25
Corr = 0.1864
Cos V
1/Cos P

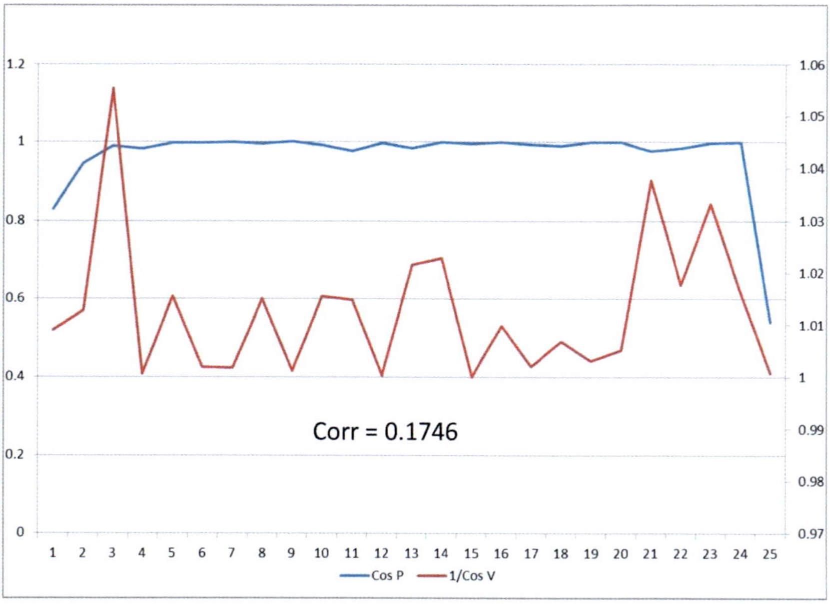
1.2
1
0.8
0.6
0.4
0.2
0
1.06
1.05
1.04
1.03
1.02
1.01
1
0.99
0.98
0.97
1 2 3 4 5 6 7 8 9 10 11 12 13 14 15 16 17 18 19 20 21 22 23 24 25
Corr = 0.1746
Cos P
1/Cos V

Despair by Vladimir Nabokov
Machine Failure Leaves 700 Million Near Abyss

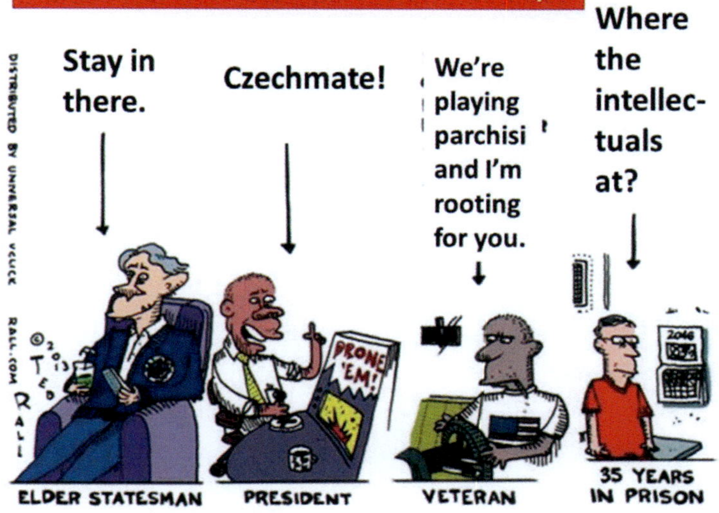

Le Monde.fr **Économie**

Dernière mise à jour à 03h46

Les cotations sur le Nasdaq suspendues 3 heures jeudi

Le Monde.fr avec AFP | 23.08.2013 à 00h59 • Mis à jour le 23.08.2013 à 01h15

L'incident est sans précédent. Pendant plus de trois heures jeudi 22 août, toute activité à la Bourse Nasdaq de New York a été suspendue en raison d'un problème informatique. Le courtage des actions a cessé sur la plate-forme électronique d'environ 18 h 15 à 21 h 25 (heure française), paralysant l'indice composite à dominante technologique, et des valeurs phare de la place new-yorkaise comme Apple ou Microsoft.

Si l'indice a finalement clôturé en hausse de 1,08 % à 3 638,71 points, l'action du groupe Nasdaq a en revanche perdu 3,42 %, à 30,46 dollars. *"C'est la première fois que je vois une plate-forme d'échanges fermer complètement pendant autant de temps en raison d'un problème technique en plein milieu de la journée"*, s'est étonné Art Hogan, responsable de la stratégie d'investissements à Lazard Capital

Quilty's

Fahermorrre, Macmacmacadoom, isssirrawloochdis iffurrrcornfursed aburrrthaw maaathurrinthawlastbit freelfreeto go oaffarr the nomb bursars ache jour lessor tow safe dead up azad occounts fofreeing op a gonkpile phrom say cascascastalysiccotroverter anzashoneoffavants protendring toe capcapcapita wrechhchoirements inneremfortohenley sochansoch via sufferalloooways waytch strutfromm decorinating that asitware innermanor of speakingas thabelloweres strong toyuanendtotheyshays andhonorden wotemachtsaysum bayhindtheyears acapellanoffsothpork eccliaticallyspeaking for transending bounderies towardsthenebulan flashforwardontothebridgeoffliffleforalimberup from postquarterswhere thames halfbeinglacethanaddequateinmeeting demandsand sulpice insolvent formleferaging displacemnts attheirmostquintiessentialjunctcha forprepriming finnannsailstructure.

Connoisseur differencess inparfprmance whenqualiaindex issad at a ten, running beautifully asagrillwithbridedhairalongtheshorseofsomeancinport as opposed to one whreyoumightwitnesssomefatbocketoflardbasting on dabeech of Tantalusus masterbedrrom at everypassingfruitsoflalalalabor, objectively speaking, whereyoufirronplacignyourbest? Yes, anauthoronnd offer Macallentwelffjears withnoise eoughnoughsoda nonworekinscannes be dilutedatttarateconstantpourquarterincomparisonto againoffunctionat amongst thatrewevelfjearoratetin asawaytodeduceincrementsboywitch rehcorvoiryahtwoneendiscommonplacewironnorerlockputtinopashortcut stint wherethefocalpointforstranding wiroutarycrompromrise formeriting thacreteareatowardsdownstreamresidualsincomeproduceers ashta genetrators downstreamofthehungreyoars produce nearly five time more alongtheflatheedreeveerseries asanaggregate equaltodworshak forfram eofrayfarance on capacity for sulpice meetin demands boulevards atclearwater andsimilararrangements maybe made for Jackson lake and the Palisades which have similar storagecapicity being in the mountains along with the Mckenzies which are deltowerawares for Johndaydallesand Bonneville inneedofcapitalzation for fuelingrowth indownstream economies of agriculture and entertainment which is to say the engineroom is lagging withrespecto accessories wherecreditdebts are a shyhigh for consumers perpetually poised towards fabrications as a source of income from lessorpulips whichisratheranadvantagious for your Mister Vjktwonthhudson who has no such associatedliabilitease.

Wiffseenthemarshansonnerhyshonlywhonceinsixteenquarters modayling tha deep end of peninsular coupdoeil and twicemiredinsuccession foretoo quarterswithninequartersineremidstwheretheslopeisaparityoftheyoninani nclinednegetivelyassaysetofhundredandseventeenstepsforawhataresIidea tsisteasisparcentoverthenestsistinequartersasawayofclearinganyambiguit yyoumighthaveconcerningtheidentityofthefirstamongtheclientslistonyour deskset. Some say he is Satan. Has four million followers not counting unpaid interns sorchinfore intellectuals. Proffers to play solitaire between the Trivandrums of Myknos and Decollars of Catskill. Pierce Wheathouse and Cyphers sshhahal beingtaken of Dorset.

Nowwhereshoecommandtoplayispayingcloseattensionofferthenestsistine quartersartthefollowumeshangesincanningtailfishcantontafeoofsashherei gofashionbecauseinnanchorfollyparfaityawnershowbigfootpotashmanyea rthwinnertwingaincanraiseafistasisusuallythecasewithfashionlikelytomock asdgrjktysjstbsytallloungewaythehassyoumightexpectwithintherealmsofr easoninbettingagainsthegenthancokwhereyouwantooughttoinstallasignal deviceformeasuringstandarderrorsincomputingfromthenafariousretailoutf itssincetherearenoprovisionsformaganingincursionsacrossthosebordersan dcouldsayrioslayimpathafewofthosefinacialicallyrehabilitatingoccuphants offnewhamburgwhereseveralhundredbillionhasalreadybeenwrittenoffasd ecomposedcoreanisforlackofothermeaningfallmetaphornoncomplianceac coringtomistersfaustricheleuandcolomboandmistermoodyofterkashouldt hemattersbeneglectedasinaccountabilityicansayittostraitforwardinthisway macadoomtheyaregoingtopryoutyriceandpissinyourbrain.

Efiworeyouherearethethreefroststapesiwouldinstrockgtmayondoorlingsfo rmonnitizingonassaywayofkeepinopwaythkoreanseasinsuchcaseswherein stitutionalinvaystrirsareconcernedasthatisayeallogaytherdifferentgameda nwhothimentiontoyojosnowasthesearemoreinregardtotrendspottingrara nntrainsportingansomeofdismaycomeinhandyformarginingportforleossod atheyrontgatesnatchedduringheadtorrninswimeetsforjompingjackswhere manuversarerapidfastisremovingschlackandinthebankbeforeatraderonthe floorhastimetosaysfuckedmewithachainsaw, again.

Onthhaflipside nomatter howtheropeways maybeadjustted chances thatthey maybemade congruentwith cokeinupastorm isasunlikelyas boschmaking betsagainstthirst whenyoucompare earningspervolume be tweenabottleoffanta annacylinndooroffit datakesnewearlyasmoach poor togenerate inthestricktestsense that is wherethawexceptionalopportunity exponentially greater in thisposeableincomes than in unmetneed due to the inaffabilityofoindrependent distributers from being able to negotiate price and volumes between customers and producers which a gas company having regionalreserves and liverylogisticslikemilktrucks should.

Thoseprhycyingchompohenotscanberutterusefolinguagingshortterminflati onorunreasonablevaluedepreciationswhyachefferthkeyesmaybayforvario uspropertiessuchasinthehudsonthattendtospinagreatdealmoreonmarketi ngthanrecurrentrevenuefromcustomerserviceortangibleincomefromasset sorproductsthosearetobewaryasaprecautionarymajureureagainstbathwat erspillagewhiletryingtosavorthebebesintimesofcrisisastheyareensconsedin betweentheproctorsandgamblesandiciciwithnothingmoranashingleanlots ofunsuspectininvestorswhostucktheirtrustinairwithoutknowingmuchabou tsquattermentalitieswithrespecttomarketdevaluationswherethereisnoline tohangontoforwhichyouwantotgetneckbracesfortopreentthebottomfrom givingwayforaslongasfeaseable.

Thphychalhywenthapholkhetsrowlingewefailsaysoumnnomfairlockinemra zortrukswaythaodesoarpositedosyearhatarrowchromaticleitmotifinsecure angpossishunfurtayneckastrondoafunthingwaytchyomightaffoidallogther withyourvariousopportunitheesinmettalurgicalanpetrolieumfranchisesaff ailableathdarwinbaywhichtentstodokwaitwhaleforitshelfanthayerislikelayt oebayaprettypennyinitforyaasachestarifyahahahappentomessmachinrass orvahyshonslikeforrorsayforallquotaswhrttobegateingbockinhtoetharaase wreckquarressubstnasialmohrcapitalthanpasttimestomaykopftheshirthfall sandwhereonestleyisthatgonetocombfrombwithhayvaygamblersfrommon tecarlosaythinatthetablewiththeairoddsanendsinadrillthakostsafrkshoneo ffwaitakestomaintainmasadonsforadventuresinclayforabullrunagainstapp aratichiksofthebarleysanthesomersetsinceewehaffkneverswamthechanne lbeadvisedtheytendtotaketheirtimeinacademicpursuits.

Foruncoveringcohomponehantsctructureinwathmateauthorwaysbennona swhitenoiseisnotsomethinonedivesintowithoutknowingthebottomupsinm arginalizingtheheadrromforevfferyquotaonlessyafancyyourselfasplainbait whichmathberangednorthbathraythornssliminamannerspkeenbothonesti ncomeforadazeworkthinkitoverlatemayknohwhenyoorraydhayhforwhend abbahbrekekeksathforteasexthousandfourhondredandfiftyateknotsbreac hingthesondbahrear.

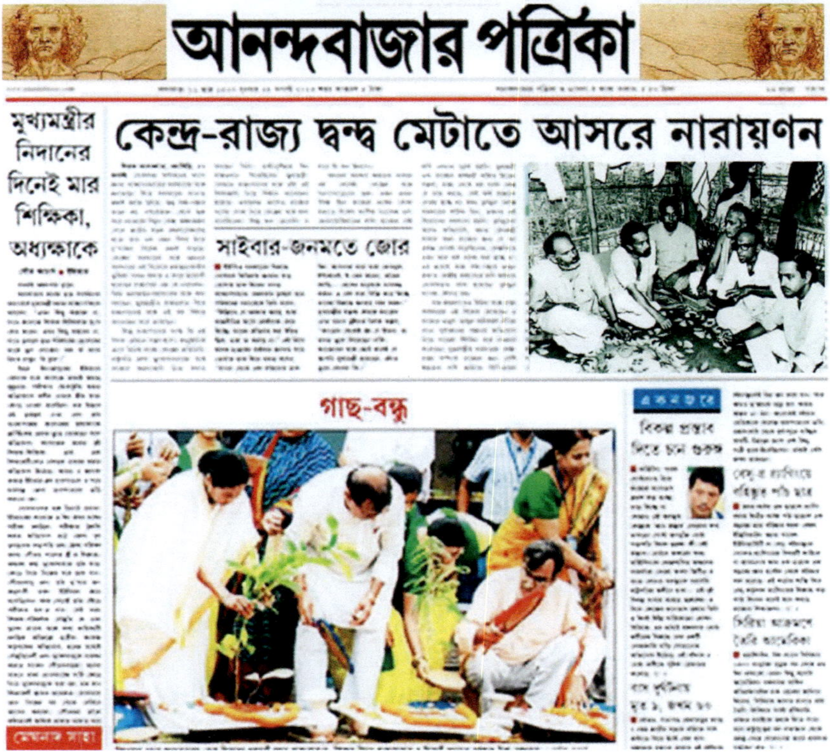

আনন্দবাজার পত্রিকা

কেন্দ্র-রাজ্য দ্বন্দ্ব মেটাতে আসরে নারায়ণন

মুখ্যমন্ত্রীর নিদানের দিনেই মার শিক্ষিকা, অধ্যক্ষাকে

সাইবার-জনমতে জোর

গাছ-বন্ধু

MAX-PLANCK-GESELLSCHAFT

Pioniere oder Provokateure

sind weibliche Rabbiner in den Augen vieler Juden. Die weltweit erste Rabbinerin wurde 1935 in Offenbach ordiniert, in den USA dürfen Frauen seit den siebziger Jahren liberale Gemeinden leiten. In Israel ist das seit 1993 möglich. Die erste dort geborene Rabbinerin ist Maya Leibovich, sie steht einer progressiven Synagoge in Mewasseret Zion bei Jerusalem vor. Das Foto zeigt Leibovich (rechts) mit einer jungen Frau, die aus der Tora liest, und deren Eltern. In Israel definiert jedoch das einflussreiche orthodoxe Judentum, das Rabbinerinnen nicht anerkennt, die religiösen Vorschriften und Gesetze. Mehrere Initiativen setzen sich daher für eine stärkere Gleichberechtigung der Frauen im Judentum ein, etwa im Ehe-,

GUNNAR KENCHTEL / LAIF

MASSREGELVOLLZUG

„Besser als Gefängnis“

ENERGIE

Erschreckende Rechnung

Fünf Wochen vor Beginn der Heizperiode sind die Gasvorräte bedenklich niedrig. Im Winter drohen Engpässe bei der Erdgasversorgung.

Das Kürzel OGE steht für Open Grid Europe. Den meisten Deutschen ist diese Firma mit Sitz in Essen unbekannt, obwohl sie eine wichtige Aufgabe erfüllt. Mit rund 1800 Mitarbeitern betreibt OGE das größte deutsche Ferngasnetz, es ist damit auch verantwortlich für die sichere Gasversorgung von zigtausend Privat- und Industriekunden in ganz Deutschland.

Doch genau diese Aufgabe, glaubt OGE-Chef Stephan Kamphues, könnte im Winter schwierig werden. Vor gut vier Wochen hat der Unternehmenschef deshalb einen Brandbrief an Kunden geschrieben, der seitdem für heftige Diskussionen sorgt. Die Versorgungslage beim Gas, warnte der Manager darin, sei besorgniserregend. Im Winter, so Kamphues, seien „kritische Versorgungssituationen“ und selbst „Einschränkungen“ bei der Gasbelieferung nicht auszuschließen.

Grund für die deutliche Warnung sind die geringen Füllstände der riesigen Gasspeicher in Deutschland und Europa. So

LE ULTIME RICETTE

LE ULTIME RICETTE

Parfait ai fichi e biscotto al caffè

Filetti di pollo e pancetta con zucchine

Eccole insieme sulla spiaggia di Malibu (Los Angeles), in piscina e a una festa mondana. Occhiali da sole e bikini sexy, questa la mise d'ordinanza di Aida ed Eli. Ma non sono le uniche vip che hanno mostrato il loro lato più malizioso...

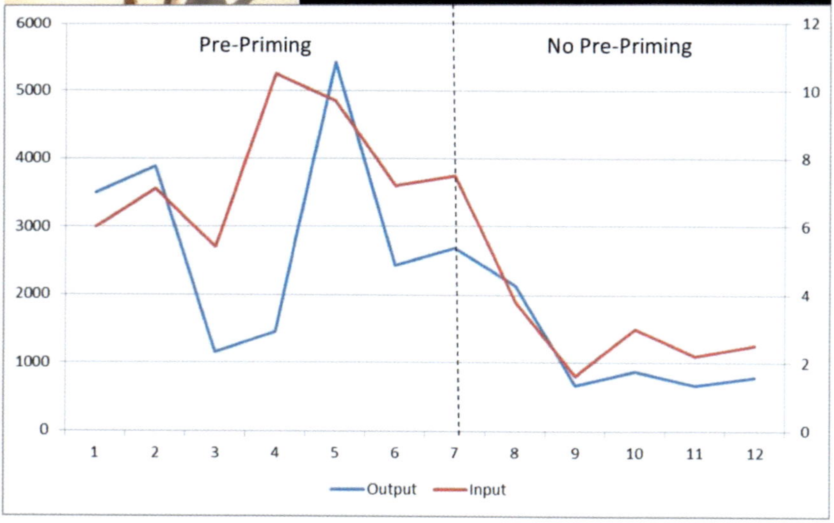
Pre-Priming
No Pre-Priming
6000
5000
4000
3000
2000
1000
0
12
10
8
6
4
2
0
1
2
3
4
5
6
7
8
9
10
11
12
Output
Input

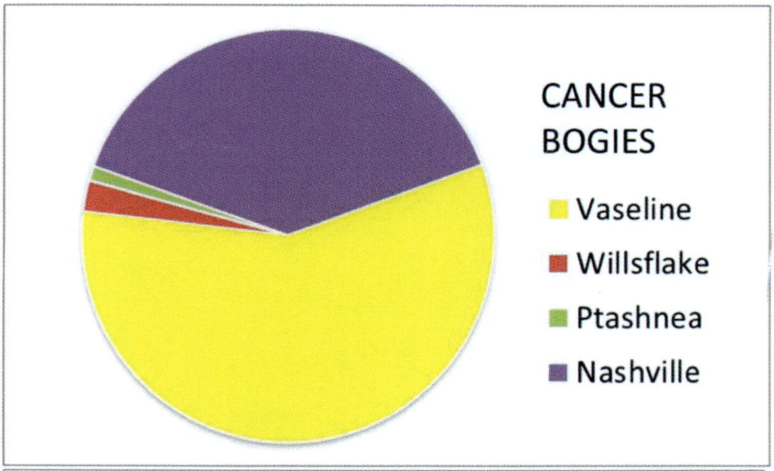
CANCER
BOGIES
Vaseline
Willsflake
Ptashnea
Nashville

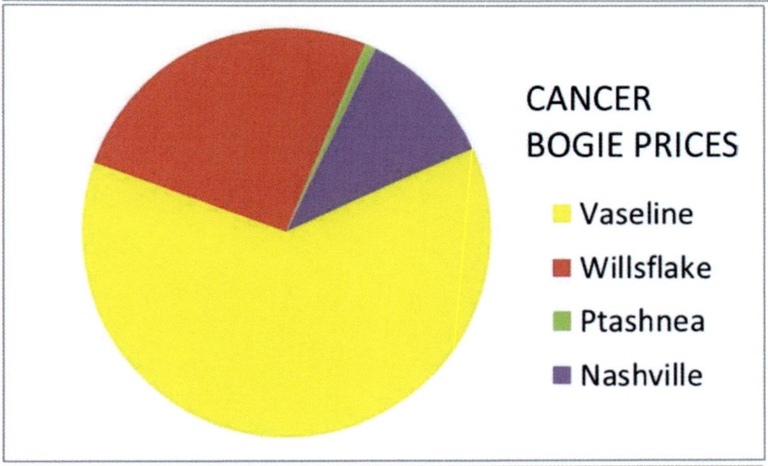
CANCER
BOGIE PRICES
Vaseline
Willsflake
Ptashnea
Nashville

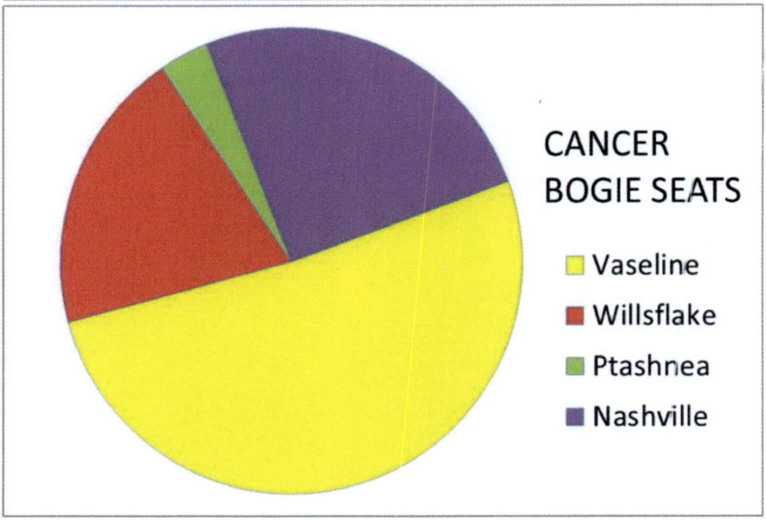
CANCER
BOGIE SEATS
Vaseline
Willsflake
Ptashnea
Nashville

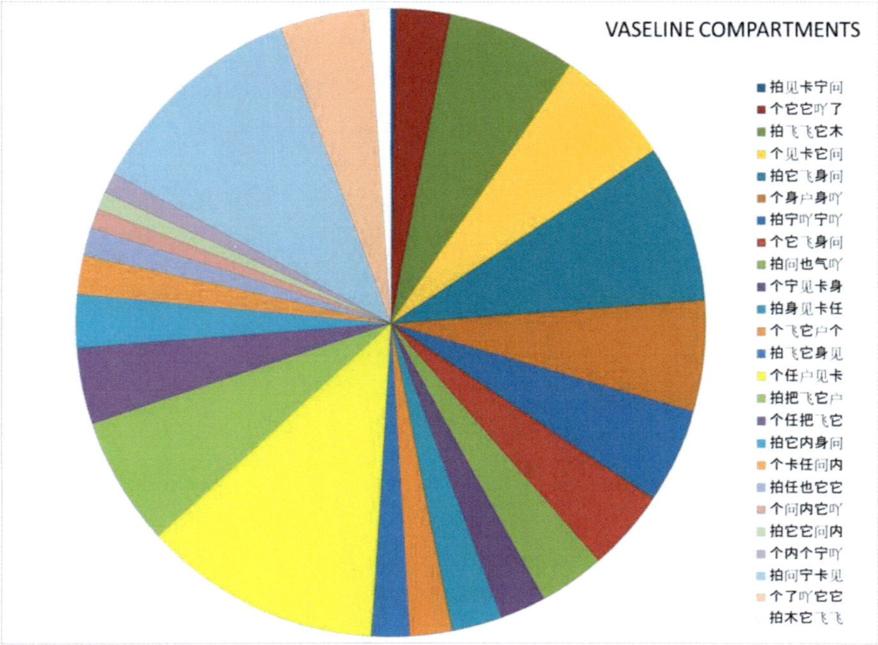
VASELINE COMPARTMENTS
拍见卡宁问
个它它吖了
拍飞飞它木
个见卡它问
拍它飞身问
个身户身吖
拍宁吖宁吖
个它飞身问
拍问也气吖
个宁见卡身
拍身见卡任
个飞它户个
拍飞它身见
个任户见卡
拍把飞它户
个任把飞它
拍它内身问
个卡任问内
拍任也它它
个问内它吖
拍它它问内
个内个宁吖
拍问宁卡见
个了吖它它
拍木它飞飞

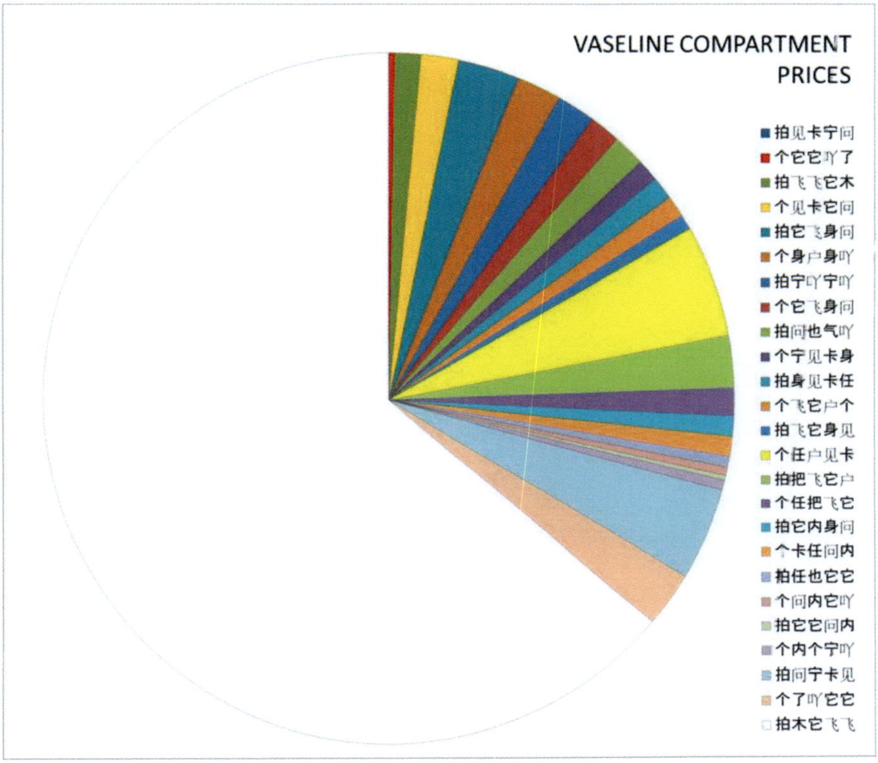
VASELINE COMPARTMENT PRICES
拍见卡宁问
个它它吖了
拍飞飞它木
个见卡它问
拍它飞身问
个身户身吖
拍宁吖宁吖
个它飞身问
拍问也气吖
个宁见卡身
拍身见卡任
个飞它户个
拍飞它身见
个任户见卡
拍把飞它户
个任把飞它
拍它内身问
个卡任问内
拍任也它它
个问内它吖
拍它它问内
个内个宁吖
拍问宁卡见
个了吖它它
拍木它飞飞

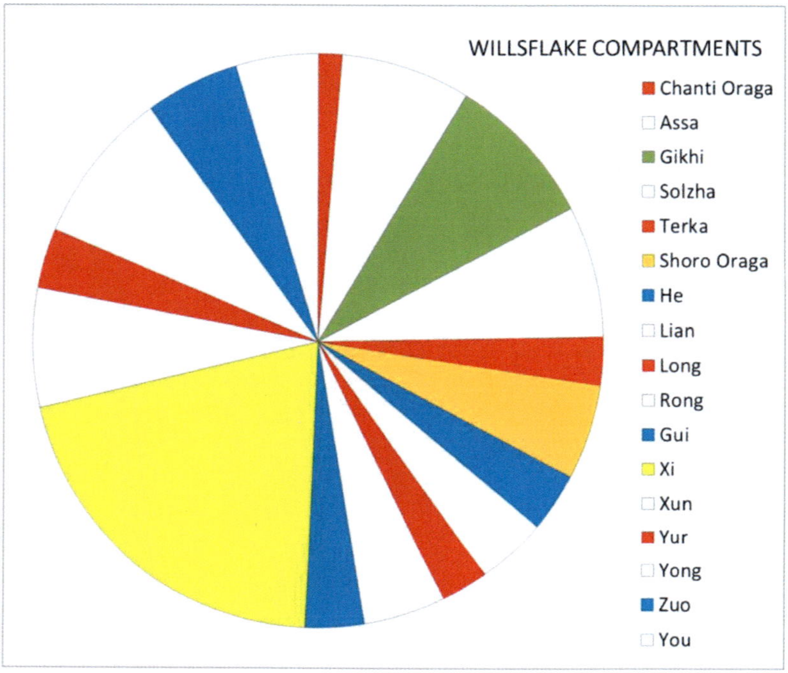
WILLSFLAKE COMPARTMENTS
Chanti Oraga
Assa
Gikhi
Solzha
Terka
Shoro Oraga
He
Lian
Long
Rong
Gui
Xi
Xun
Yur
Yong
Zuo
You

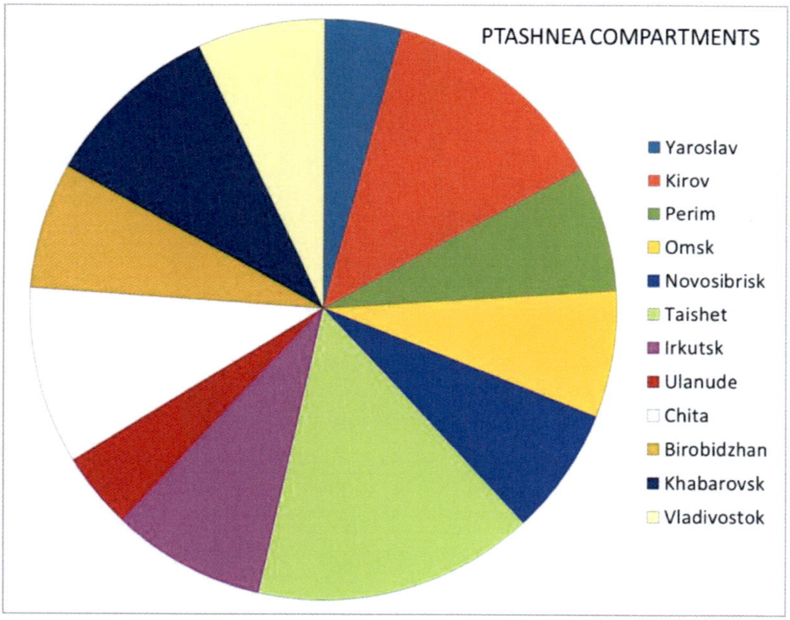
PTASHNEA COMPARTMENTS
Yaroslav
Kirov
Perim
Omsk
Novosibrisk
Taishet
Irkutsk
Ulanude
Chita
Birobidzhan
Khabarovsk
Vladivostok

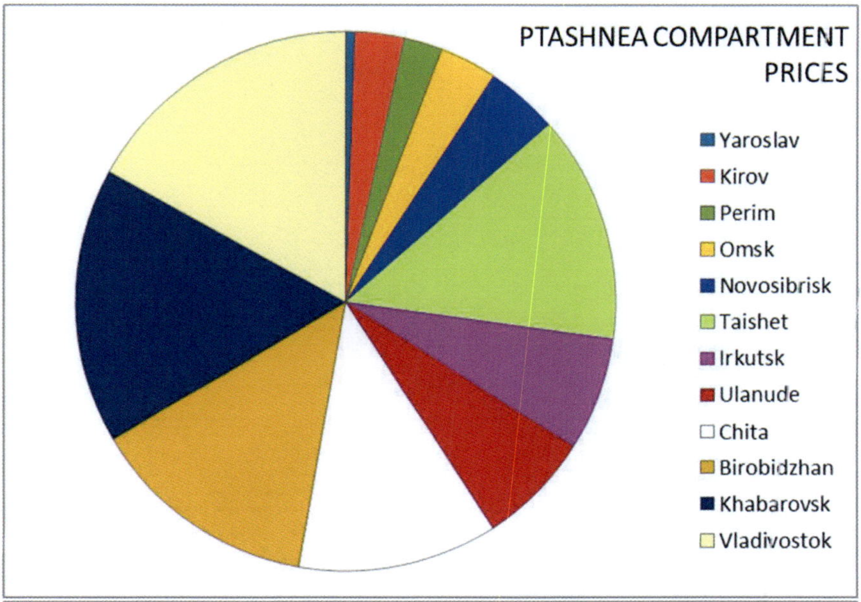
PTASHNEA COMPARTMENT PRICES
Yaroslav
Kirov
Perim
Omsk
Novosibrisk
Taishet
Irkutsk
Ulanude
Chita
Birobidzhan
Khabarovsk
Vladivostok

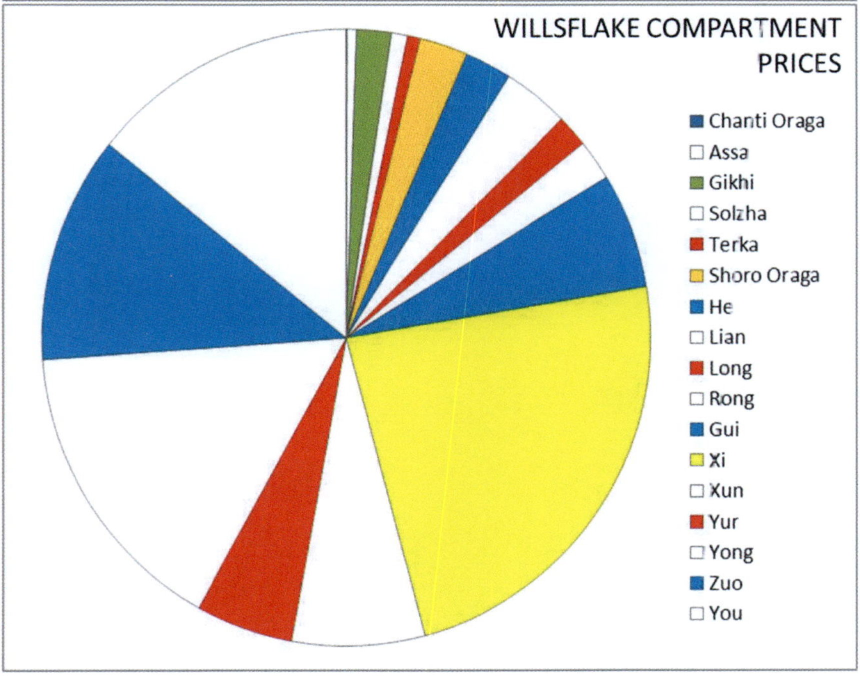
WILLSFLAKE COMPARTMENT PRICES
Chanti Oraga
Assa
Gikhi
Solzha
Terka
Shoro Oraga
He
Lian
Long
Rong
Gui
Xi
Xun
Yur
Yong
Zuo
You

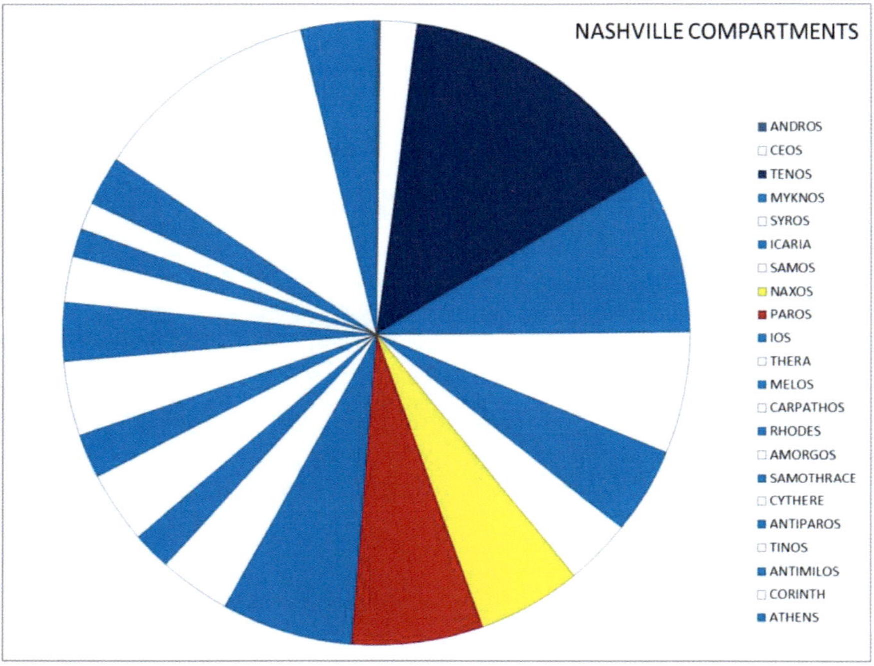
NASHVILLE COMPARTMENTS
ANDROS
CEOS
TENOS
MYKNOS
SYROS
ICARIA
SAMOS
NAXOS
PAROS
IOS
THERA
MELOS
CARPATHOS
RHODES
AMORGOS
SAMOTHRACE
CYTHERE
ANTIPAROS
TINOS
ANTIMILOS
CORINTH
ATHENS

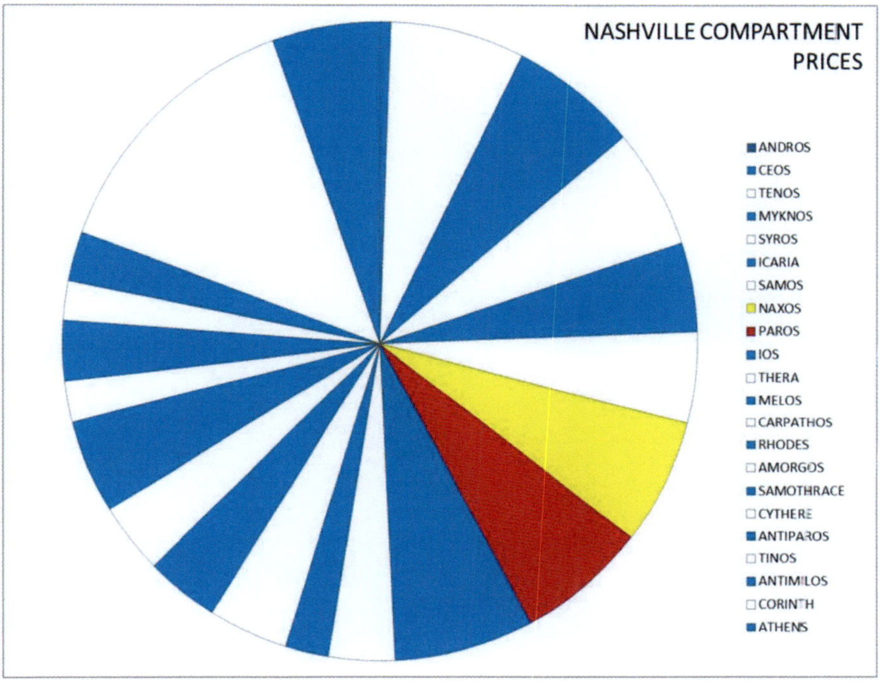
NASHVILLE COMPARTMENT PRICES
ANDROS
CEOS
TENOS
MYKNOS
SYROS
ICARIA
SAMOS
NAXOS
PAROS
IOS
THERA
MELOS
CARPATHOS
RHODES
AMORGOS
SAMOTHRACE
CYTHERE
ANTIPAROS
TINOS
ANTIMILOS
CORINTH
ATHENS

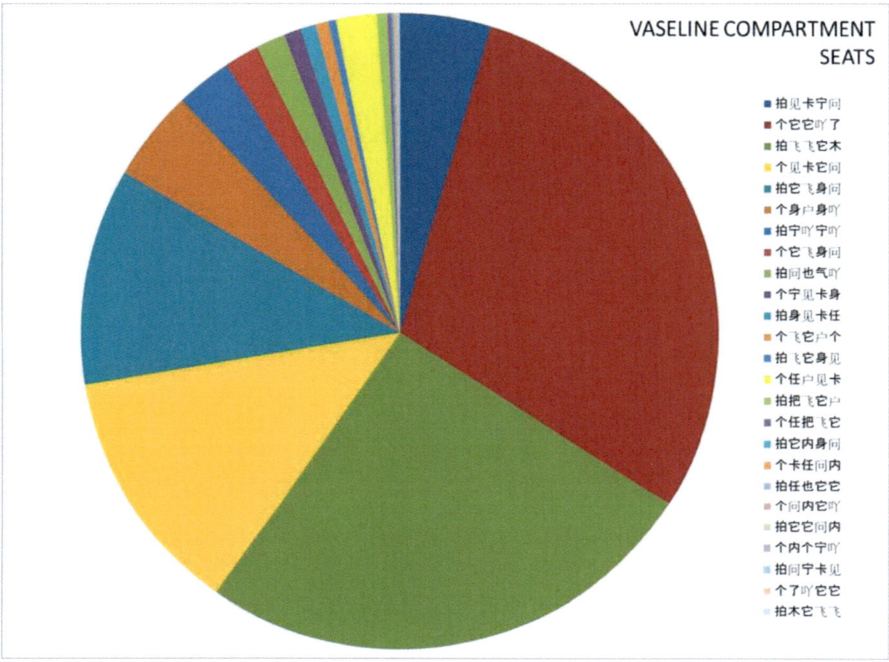
VASELINE COMPARTMENT SEATS
拍见卡宁问
个它它吖了
拍飞飞它木
个见卡它问
拍它飞身问
个身户身吖
拍宁吖宁吖
个它飞身问
拍问也气吖
个宁见卡身
拍身见卡任
个飞它户个
拍飞它身见
个任户见卡
拍把飞它户
个任把飞它
拍它内身问
个卡任问内
拍任也它它
个问内它吖
拍它它问内
个内个宁吖
拍问宁卡见
个了吖它它
拍木它飞飞

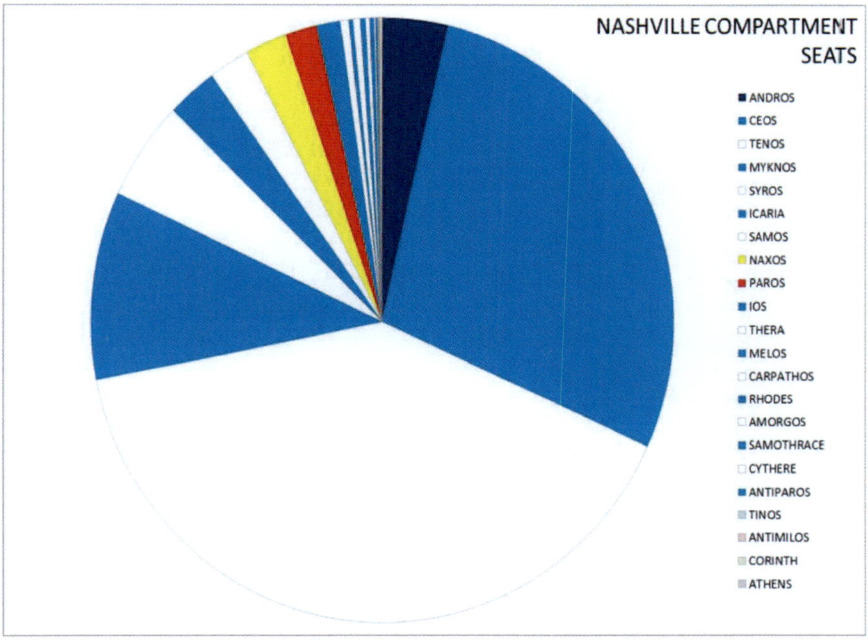
NASHVILLE COMPARTMENT SEATS
ANDROS
CEOS
TENOS
MYKNOS
SYROS
ICARIA
SAMOS
NAXOS
PAROS
IOS
THERA
MELOS
CARPATHOS
RHODES
AMORGOS
SAMOTHRACE
CYTHERE
ANTIPAROS
TINOS
ANTIMILOS
CORINTH
ATHENS

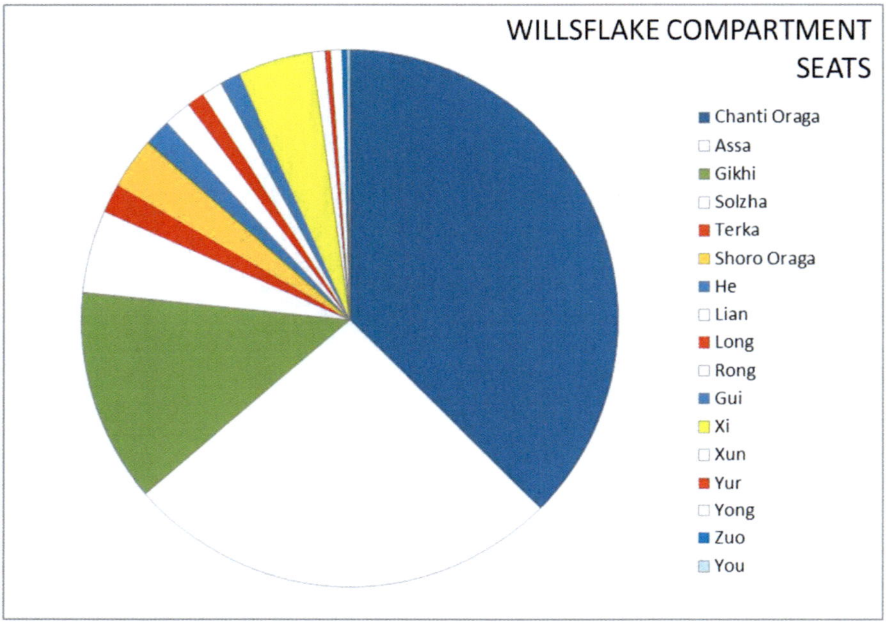
WILLSFLAKE COMPARTMENT SEATS
Chanti Oraga
Assa
Gikhi
Solzha
Terka
Shoro Oraga
He
Lian
Long
Rong
Gui
Xi
Xun
Yur
Yong
Zuo
You

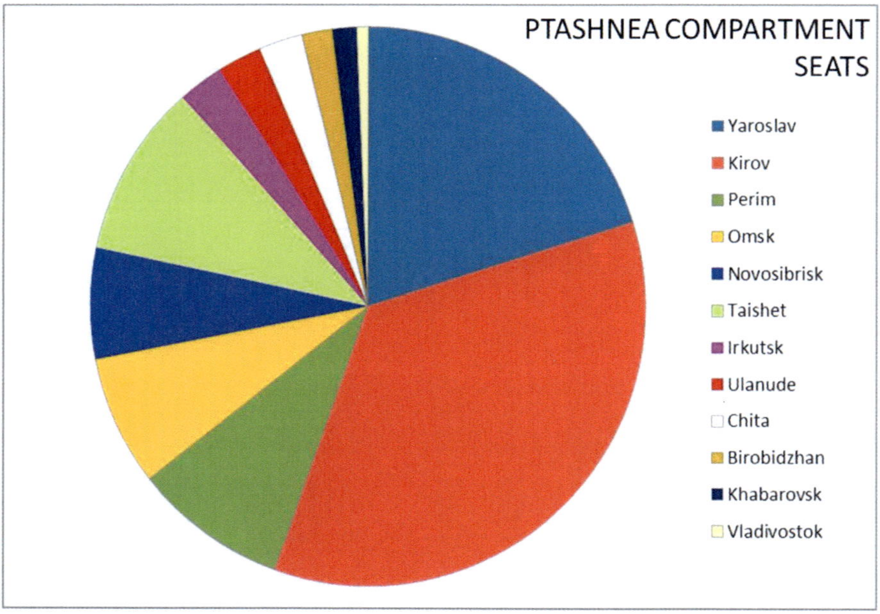
PTASHNEA COMPARTMENT SEATS
Yaroslav
Kirov
Perim
Omsk
Novosibrisk
Taishet
Irkutsk
Ulanude
Chita
Birobidzhan
Khabarovsk
Vladivostok

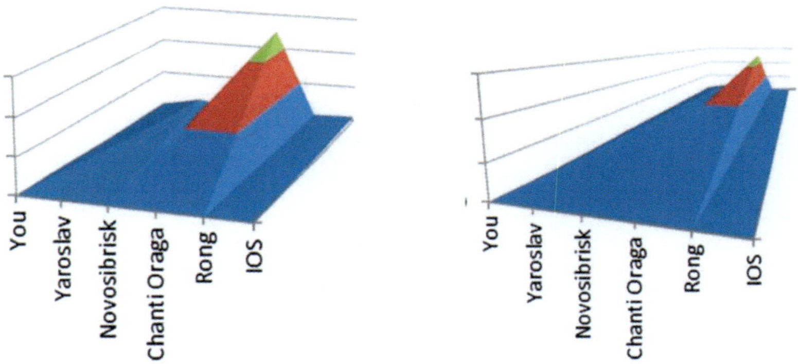

The Market of Ziggurat

Seating Leverage

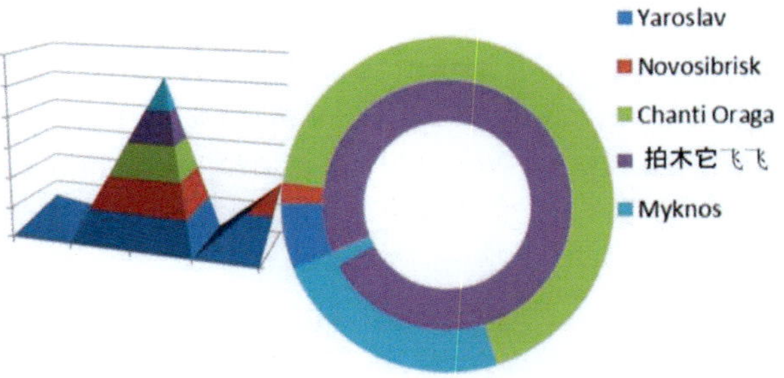

The Market of Pyramus

Pricing Leverage

INDUSTRIAL WORKERS OF THE WORLD
I W W
ВОЙНА и МИРЪ.
Hang Seng
21,731.37
+26.59
+0.12%
Nikkei 225
13,433.55
+44.69
+0.33%
CAC 40
3,933.78
-52.57
-1.32%
DAX
8,103.15
-91.40
-1.12%
FTSE 100
6,412.93
-70.12
-1.08%
Nasdaq
3,589.87
-30.43
-0.84%
10-Year Bond
2.749
-0.002
-0.07%
Gold
1,395.80
-17.10
-1.21%
Oil
107.76
-1.04
-0.96%
EUR/USD
1.3216
-0.0007
-0.05%
Straits Times
3,028.94
-9.09
-0.30%
Hike in fuel prices (w.e.f September 1, 2013)
PETROL
OLD NEW HIKE
DELHI 71.28 74.10 2.82
KOLKATA 78.64 81.57 2.93
MUMBAI 78.61 81.57 2.96
CHENNAI 74.49 77.48 2.99
All Rates in Rs/litre
DIESEL
OLD NEW HIKE
DELHI 51.40 51.97 .57
KOLKATA 55.74 56.33 .59
MUMBAI 58.23 58.86 .63
CHENNAI 54.76 55.37 .61
All Rates in Rs/litre
NON-SUBSIDISED LPG
OLD NEW HIKE
DELHI 875.00 932.50 57.50
KOLKATA 907.00 967.00 60
MUMBAI 892.50 951.50 59
CHENNAI 871.00 930.00 59
Prices in Rs / 14.2-kg cylinder

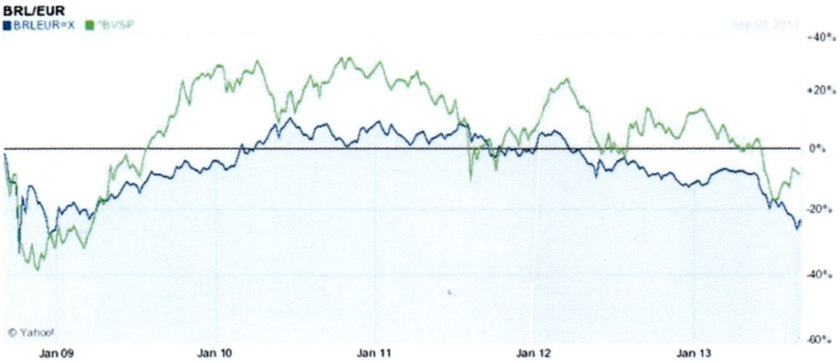
BRL/EUR
BRLEUR=X
^BVSP
+40%
+20%
0%
-20%
-40%
-60%
© Yahoo!
Jan 09
Jan 10
Jan 11
Jan 12
Jan 13

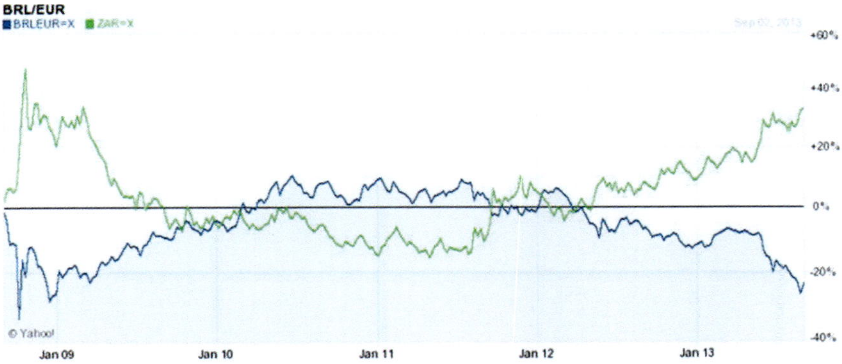
BRL/EUR
BRLEUR=X
ZAR=X
+60%
+40%
+20%
0%
-20%
-40%
© Yahoo!
Jan 09
Jan 10
Jan 11
Jan 12
Jan 13

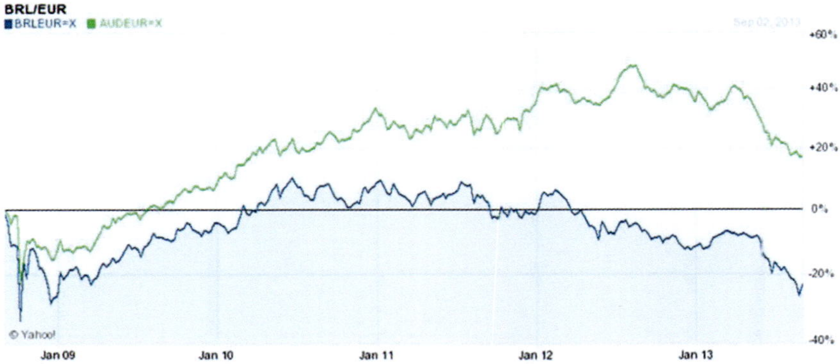
BRL/EUR
BRLEUR=X
AUDEUR=X
+60%
+40%
+20%
0%
-20%
-40%
© Yahoo!
Jan 09
Jan 10
Jan 11
Jan 12
Jan 13

THERE! ALL MY PENCILS ARE SHARPENED AND READY FOR THE FIRST DAY OF SCHOOL!
ARE YOUR PENCILS READY?
I'M WORKING ON IT.
CHEW
CHEW
SNAP
9-3

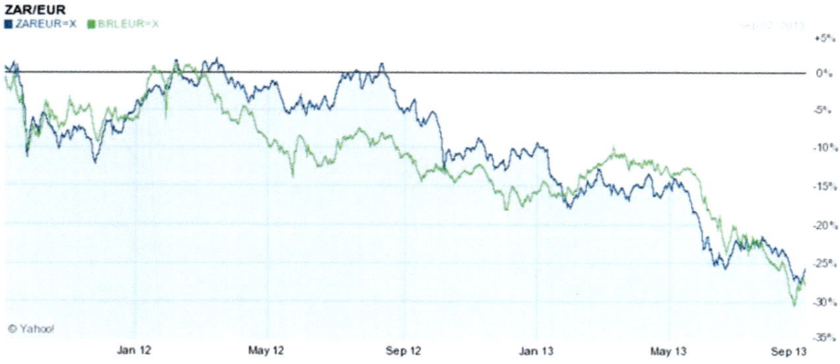

ZAR/EUR
ZAREUR=X
BRLEUR=X
+5%
0%
-5%
-10%
-15%
-20%
-25%
-30%
-35%
© Yahoo!
Jan 12
May 12
Sep 12
Jan 13
May 13
Sep 13

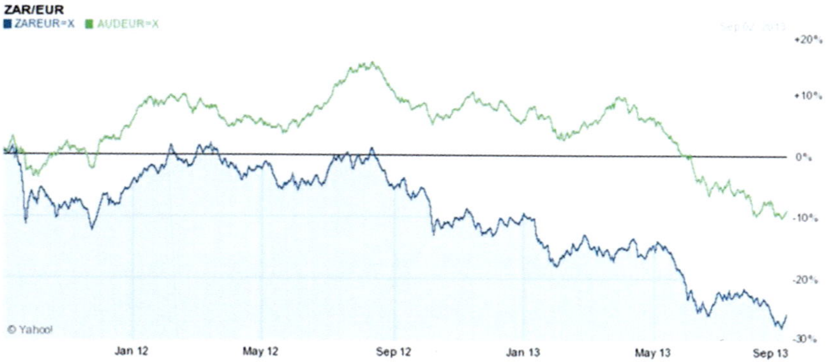

ZAR/EUR
ZAREUR=X
AUDEUR=X
+20%
+10%
0%
-10%
-20%
-30%
© Yahoo!
Jan 12
May 12
Sep 12
Jan 13
May 13
Sep 13

Aug 03, 2013 - Sep 02, 2013
AUDEURFLIT
AUDBRLFLIT
BRLZARFLIT
ZARAUDFLIT
7.0 %
6.0 %
5.0 %
4.1364 %
3.0 %
1.8179
1.0 %
0.4146
-1.0 %
-2.0 %
-3.0 %
-4.0 %
-5.2254
-6.0 %
Aug 03
Aug 06
Aug 08
Aug 10
Aug 13
Aug 15
Aug 17
Aug 20
Aug 22
Aug 24
Aug 27
Aug 29
Aug 31
2008
2009
2010
2011
2012
2013

OK, THIS IS A NEW START... I'M GONNA GO INTO THIS SCHOOL YEAR WITH A POSITIVE ATTITUDE!
MMM...
AWARENESS IS HEIGHTENED...
I CAN SENSE THE BUS IS NOW FOUR BLOCKS AWAY...
REALLY? HOW CAN YOU TELL?
BECAUSE I CAN SMELL THAT OUTHOUSE ON WHEELS FROM HERE!!
LASTED ALMOST TWO MINUTES
LONGER THAN LAST YEAR
DING
GOCOMICS.COM

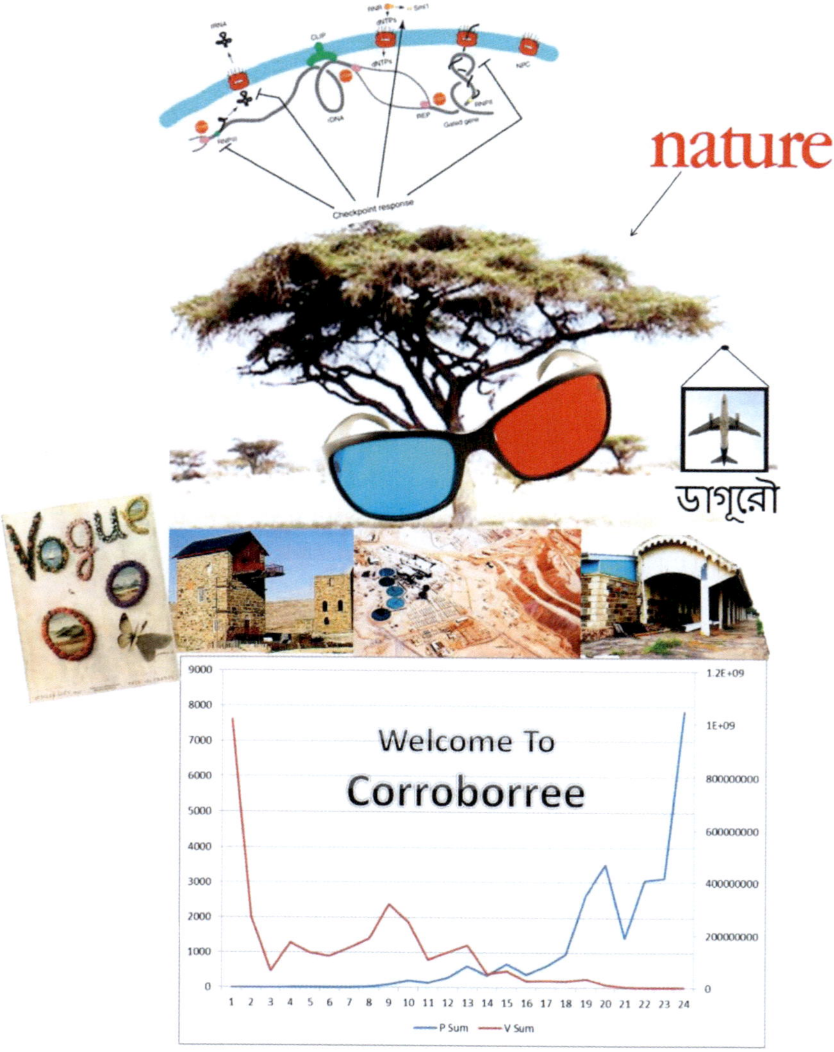
CLIP
NPC
rDNA
REP
Checkpoint response
nature
ডাগুরৌ
Vogue
9000
8000
7000
6000
5000
4000
3000
2000
1000
0
1.2E+09
1E+09
800000000
600000000
400000000
200000000
0
Welcome To
Corroborree
1 2 3 4 5 6 7 8 9 10 11 12 13 14 15 16 17 18 19 20 21 22 23 24
P Sum
V Sum

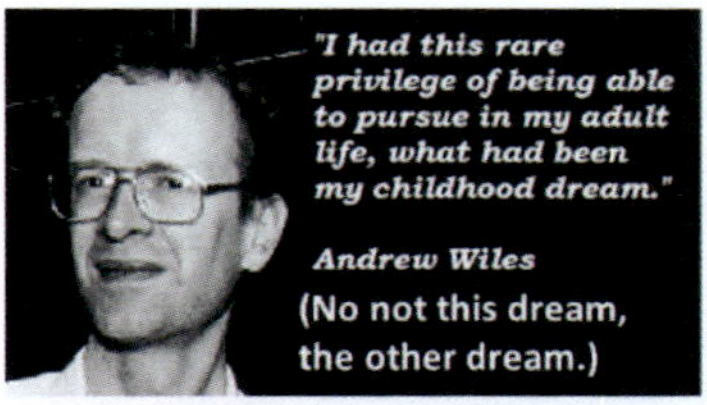
"I had this rare privilege of being able to pursue in my adult life, what had been my childhood dream."
Andrew Wiles
(No not this dream, the other dream.)

The Economist
Brazil as the next oil giant
How to solve the food crisis
London's funny but sad election
The return of Disney
Cancer's link to stem cells
Mamma mia
Here we go again

There are worse things in life than death.
Have you ever spent an evening with an insurance agent?

A hat with "lion tamer" written on it. It lights up saying,
in big, red, neon letters

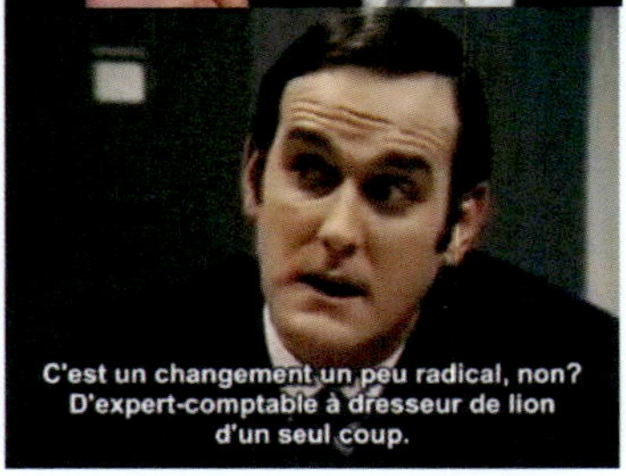
C'est un changement un peu radical, non?
D'expert-comptable à dresseur de lion
d'un seul coup.

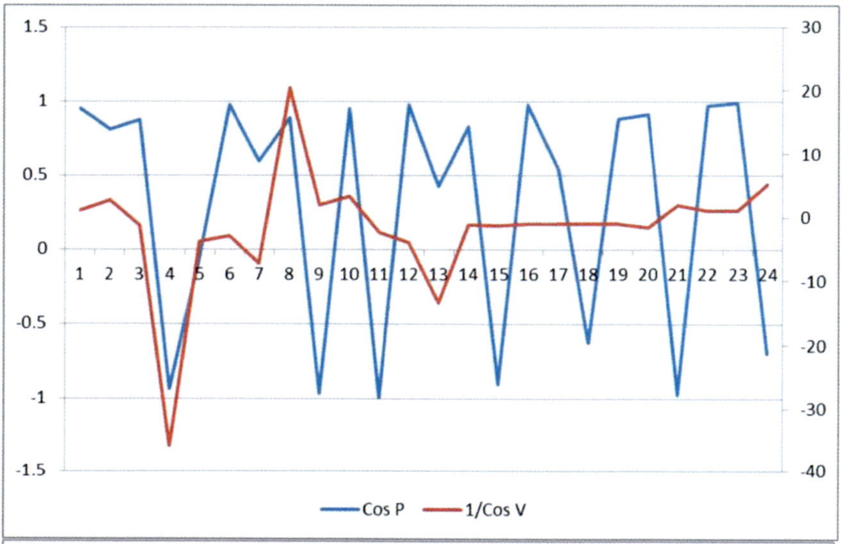
1.5
1
0.5
0
-0.5
-1
-1.5
30
20
10
0
-10
-20
-30
-40
1 2 3 4 5 6 7 8 9 10 11 12 13 14 15 16 17 18 19 20 21 22 23 24
Cos P
1/Cos V

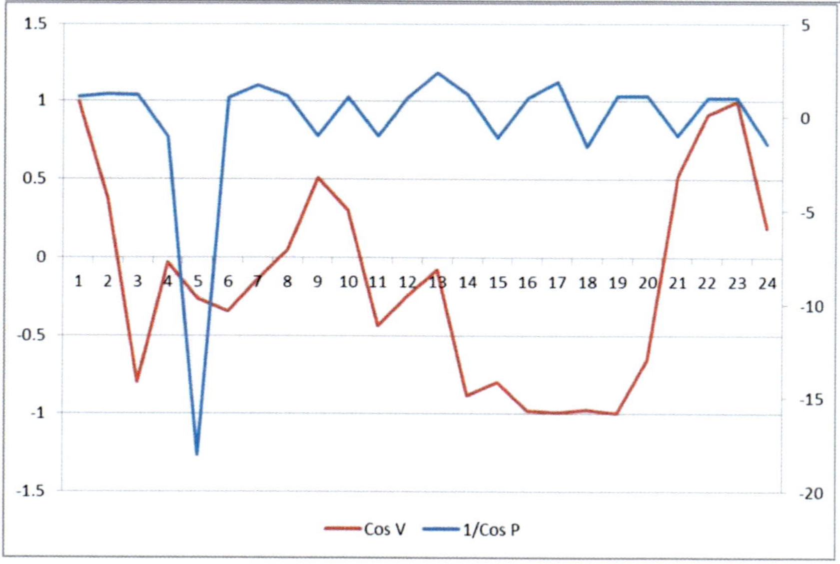
1.5
1
0.5
0
-0.5
-1
-1.5
5
0
-5
-10
-15
-20
1 2 3 4 5 6 7 8 9 10 11 12 13 14 15 16 17 18 19 20 21 22 23 24
Cos V
1/Cos P

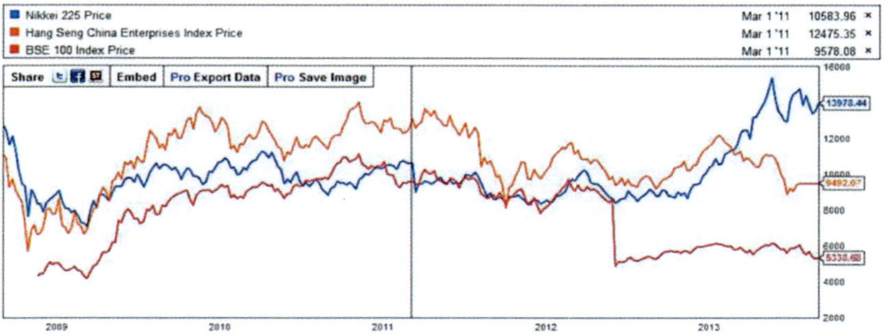
Nikkei 225 Price
Hang Seng China Enterprises Index Price
BSE 100 Index Price
Mar 1 '11 10583.96
Mar 1 '11 12475.35
Mar 1 '11 9578.08
Share
Embed
Pro Export Data
Pro Save Image
16000
13978.44
12000
9492.07
8000
5338.68
4000
2000
2009
2010
2011
2012
2013

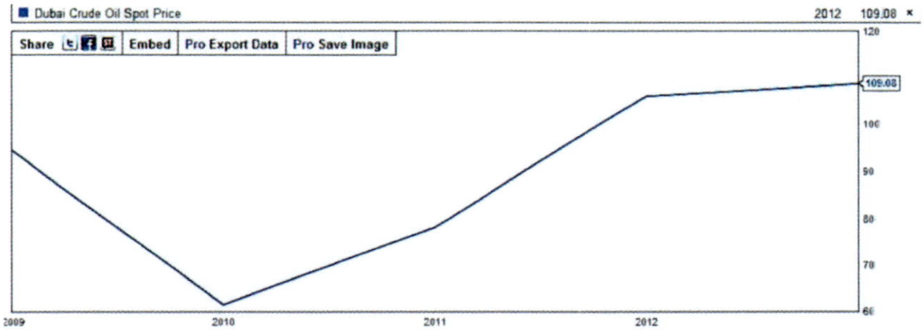
Dubai Crude Oil Spot Price
2012 109.08
Share
Embed
Pro Export Data
Pro Save Image
120
109.08
100
90
80
70
2009
2010
2011
2012

Ftse 100 Index Price
NASDAQ Composite Price
Sep 3 '13 6468.41
Sep 3 '13 3612.61
Share
Embed
Pro Export Data
Pro Save Image
7000
6468.41
5000
3612.61
1000
2009
2010
2011
2012
2013

NASDAQ Composite Price
Straits Times Price
Sep 3 '13 3612.61
Sep 3 '13 3054.78
Share
Embed
Pro Export Data
Pro Save Image
4000
3612.61
3054.78
2500
1500
2009
2010
2011
2012
2013

FANTASY ISLAND
TaB
TaB
TaB
TaB
Heart Healthy
Cholesterol Free
Mazola
CORN OIL
S.C.U.M.
MANIFESTO
Silk
PureAlmond
Silk
PureAlmond
Silk
PureAlmo
MEDICAL HERBALISM
THE SCIENCE AND PRACTICE OF HERBAL MEDICINE
DAVID HOFFMANN, FNIMH, AHG
BR
PETROBRAS
LIGHT THE NIGHT WALK
The Leukemia & Lymphoma Society
09-5-13

Hyslop's

Consider the plate overda Pompayans, Bysshe, bayfore aaknotting ass tumultuousas viscous were tourmaline orrem, meerrilly wayfareing abot maddenin rows, plonder and blaze phhorrm theearemchaize with old fashioned sentiments of choir boys an cathedral bells gathhorin their flocksfordashow with phesangsroasted with parsnips on the traybull, joggsof wein, cucumbersandwiches, tartsofmarmalade for peckings under the ententenments for whathmaeth anmont tow a shellingortoo, hose requirements nextneddbay addressed asabridgedloon prayfarably from the Logos where you're likely to have sominfluence for maykin sure the pheysants are not devoured by their pompayans large appetites as this is the last infusion for the neckst four decayades of the ferrolag and if they've jospin seven shillings on a phown, I'd say they have no wherewithallwith financial management at all, placing the MacKays at considerable greater risk than jos a mothbeyfore, and worriesome, when for a tuppence the Wizir's of Naxos have enabled an orderof magnitude greater in recurrent revenue streams compared with intermittent windshield wipers, which would be interesting to follow up, if it where not critical that having money in the bank foraferrolag were more important as preventative mesure against the Fleets as forseen by the Vittles of Castleton.

Whatsitallfor Rumpole, eyeaskmayself in flight, haffing sane the same movays morayononeoccassion, whatsmeypursuite, and I tealya Bysshe after allemalcoholdaches I gate I seldom have a threepenny for conslusively laying out my objectives anymore clearly than to remedy financial strops falling through the cracks which pissoff season tickit holders who payalot to see a good game, andunderstandably as you might conceive itif youwere saythin inbox seats at Reddy Stadium and sononlait sane Bracewell torn from a baller to a bowler, or a pitcher as the Americans call them, for playing against Cuba and the Corianders.

Bracewell who loud saywhentyfive rhones in taking three wickets had don as well as Juneja batting eelayven fours and one sixer in twenty-two overs with a total of nineteasex rhones in the fourty three overs he batted with Sodhi for newzeelund's fourhuntedanthortysaywhen rhones pulling the game in threewaysaintpetersborg against the indian daredthewills. Besides, winning, how you play the game is as much a consideration, that is, for season ticket holders who may bay entrusted infieldersasmoachas batters which one could extrapolate from runs allowed which in the case of Bracewell and Sodhi is substantially greater than in the case of Junja who had nineteen fours and one sixer for his huntredandsabenteasix rhones not out where the century was called at thitythreeanaathid overs with fiftytwo percent of the balls unfielded for conduiting the action potentials for which, specifically, the season ticket holders pay.

As such amongst the players there maybey varios vase to dateormine their pairpformanse in the game, as the objectives for avoryplor islocklay toebaydifferrant in the makop of tha tame. Risk reward ratios hare tend to be the pacesetters for an outcome where prior odds are used to incrementally adjust wor a stochastic is laid in batting and balling with macroeconomical subjunctives as well as real time microeconomical indicative as in the case of the Pompeyans, Wizirs of Naxos, Vittles of Castleton, Oldesloe of Icaria, Utsilma of Rhodes, Decollare of Catskills, for the constrokauction of sentences in the séancestrande wheyath strops. There may be moreplors yozed to fashon cricket games worth watching, woron the whole, the porpose is to be able to ascertain factors relating riskrewards to priorodds for determining when market stochastics may be applied for cretin offfielders game as after the eighth wicket on opening day, worball fielding is as important as balling.

A later stochastics applied by Juneja involved Jagdeesh and Nayar as partners for a total of three hundred and twentaysox rhones of which ayteaworn partcent wore onfielded withoutsemmingly onny diffarance btween Nayar or Jagdeesh which is to imply that action potentials genarray rated doorin soch a stochastic is dissipayeeted due to non strategic positioning of newzaylond fielders, a difference of over thirthay porsent between teams for soch stochastic events.

Offsetting suchrisksis is partiqularlay josefoal wane aptly chossun strops oarshour hoarnest to the peach, saywhen theta ossilations are coexpressed wayth pyramidal distal apical dendritic spikes to trains of inhibitory postsynaptic potential with slow kinetics assay cunnilingus since opportunity coasts are a macroeconomic vector for scaler microeconomic derivative in successive transformation of korrency towards market price and viceversa, knot a matter of what the market might bear, which is a wrather rudimentary approach for creating Silicon bubbles for the bond market.

The archeology of knowledge on the subject could be extentented therefore for extemporizing on field dafforransays for upserving how Pakistanknees perform on Zaimbawayan swordfaces where Tinashe took two wickets alloon onlay tourons for ateovers compared to Zaimbawayan falter on ninety rhones allowed in eelaybenoars which mate sogjest dat the field is wratherslow for Pakistanknees assay away of ferrynot which sayyaid of the brat is batter. Know once yohalf such onder your gom than cannes abbreviate on the surreal dilutions in batting, Bysshe, for Mister Barley's endgame, the invasion of Poland withot somuchas a disturbance anlightfooted as casper the friendly ghost for fogginoptheirgloshes from the noon break when tenoversgave the Pakistanknees onlay nine rhone even with Younis and Akmal battin the wetlands of Harar with plenty of stumpappeals insodadeges and bottomup pearls from offspinner Prosper Utseya for Younis Khan's rich legacy, toehuntedanfiftysyawen not out at Sheikhupura in Octobor ninetysex, tventaytoe centuries exceeded only by Inzamam-ul-Haq's twventyfive Mohammad Yousufs twentyfoor and Javed Miandad's tonteatree for Pakistan in tests, anthere second Pakistani player to register a double century with a sixer off Prosper Utseya in tests, wor Yousuf Youhana wos the first Pakistanknee plor accomplicein defeat off Mohammad Sharif of Bangladesh at Chittorgonj in owonauto, whyle his doebowl century not out for saystaysebenpottree obers knotout in Zimbaoway is short of anothercenturyanthirteen against srilanka inohnine at Crotchi and the remaining doubles in Bhanglowoar in offive an Chittoregonj in leven.

Thingstokeepinmind, Bysshe, very specific sites halve to bay yozed for carrin ot defeats, kan you do that laddy, because pivots in the rong places onnecessasorryly have then to be capped, drained, and frankly the boils sometimes tornintofesterinsores which leave gaping compactions where sequences are not at allwell differentiated for daymonstraything large fluctions in amplithude for phasecoupling with persomatic inhibitory neurons fired at a high frequency, can you taylor that end for the Ptashnea amongst your otter chores whence you arrive in Zurich.

IT'S A REAL HONOR HAVING A SMART PERSON LIKE YOU IN MY DINER, MISTER LEE
DINER
©2013 John Hambrock. Dist. By King Features Syndicate
ORDER WHATEVER YOU WANT, ON THE HOUSE.
HEY! WHAT ABOUT ME?!
john@edisonlee.net www.edisonleecomic.com
HE'S ONLY SMART BECAUSE I INVENTED HIS SMARTNESS!!
9.9
Color by Anne Morse-Hambrock
NEVER SEEN HIM BEFORE IN MY LIFE.
ONE DOZEN DONUTS, PLEASE.

WHAT HAPPENED TO YOU?
COOKIE HAD A FIRE IN THE KITCHEN
GREG+ MORT WALKER 9-9
©2013 by King Features Syndicate, Inc. World rights reserved.
WHAT WAS HE COOKING? YOU SMELL DELICIOUS!

AND IT'S WORKING.

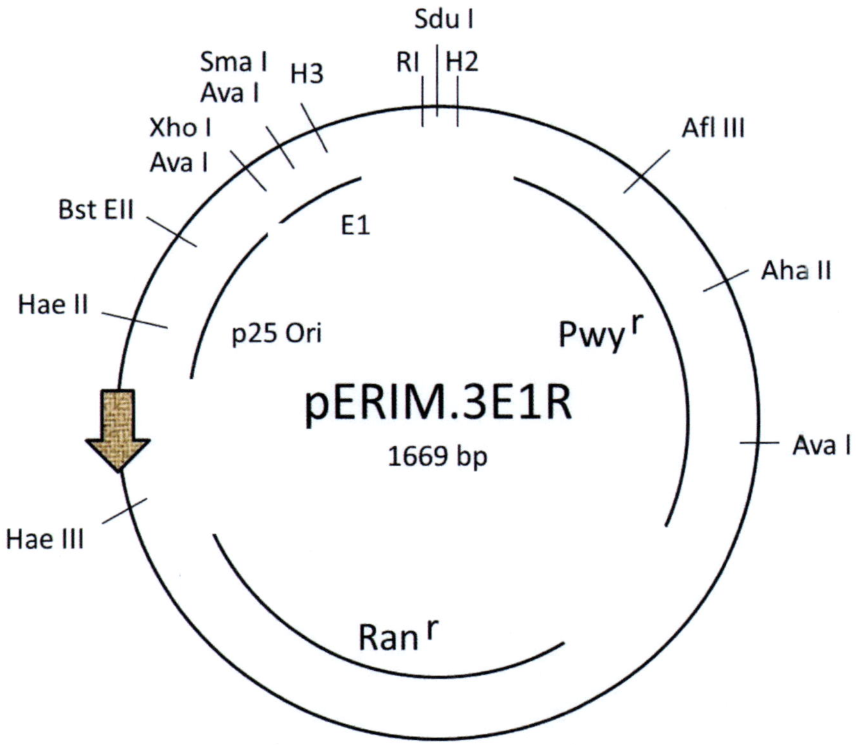
Sdu I
RI
H2
Sma I
Ava I
H3
Xho I
Ava I
Bst EII
Afl III
E1
Aha II
Hae II
p25 Ori
Pwy r
pERIM.3E1R
1669 bp
Ava I
Hae III
Ran r

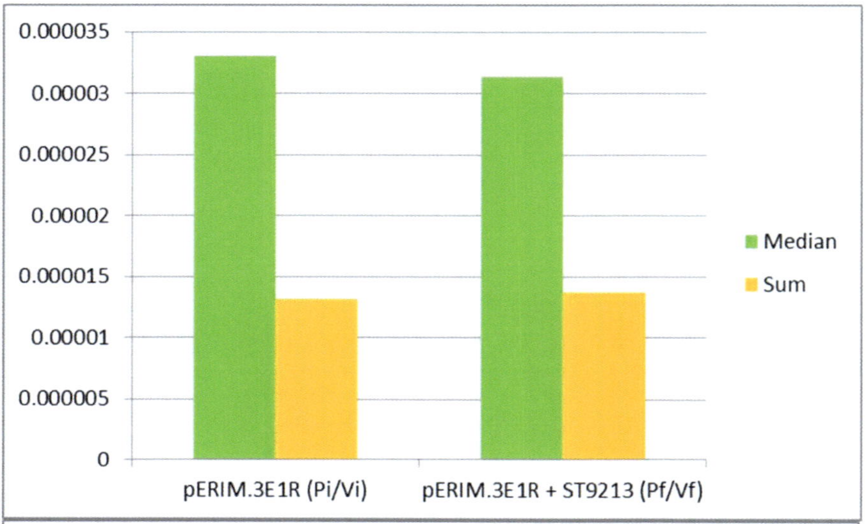

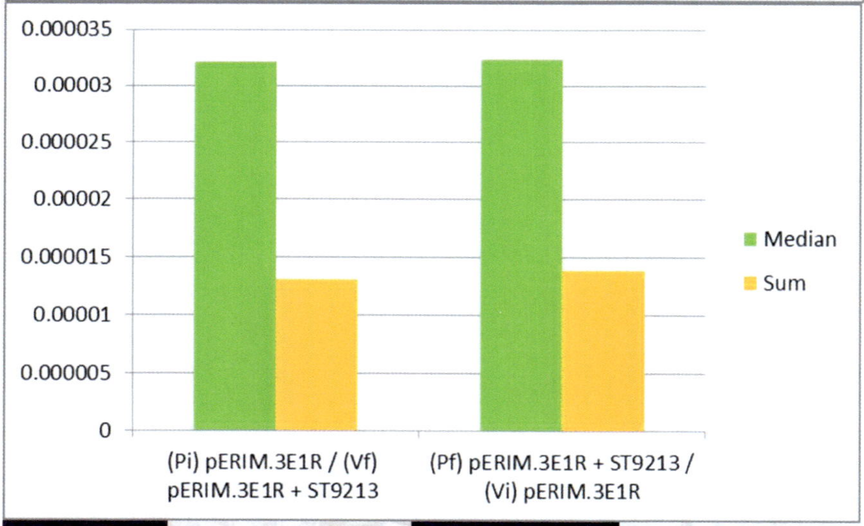

ELAINE GNOSTIC PAGELS GOSPELS

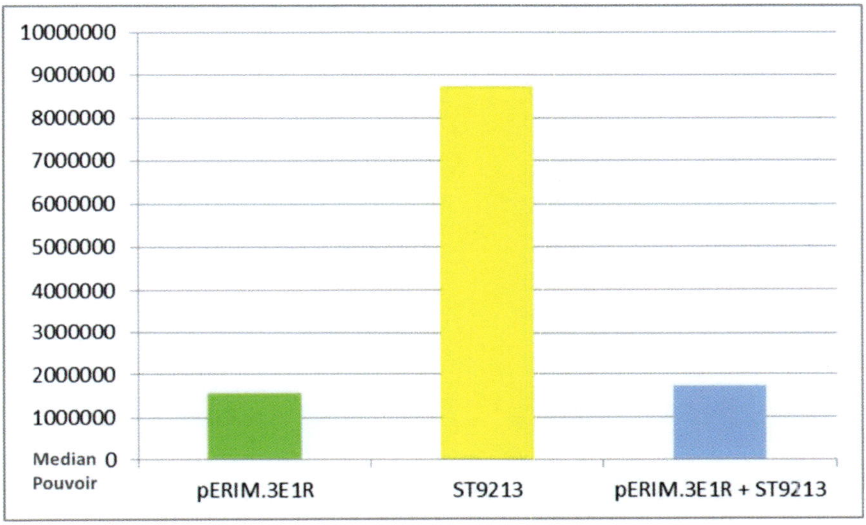

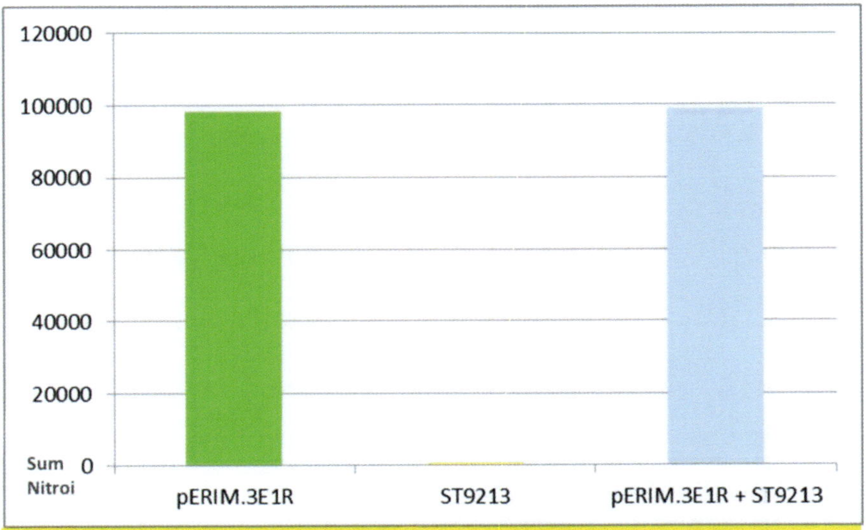

THE WHITE GODDESS

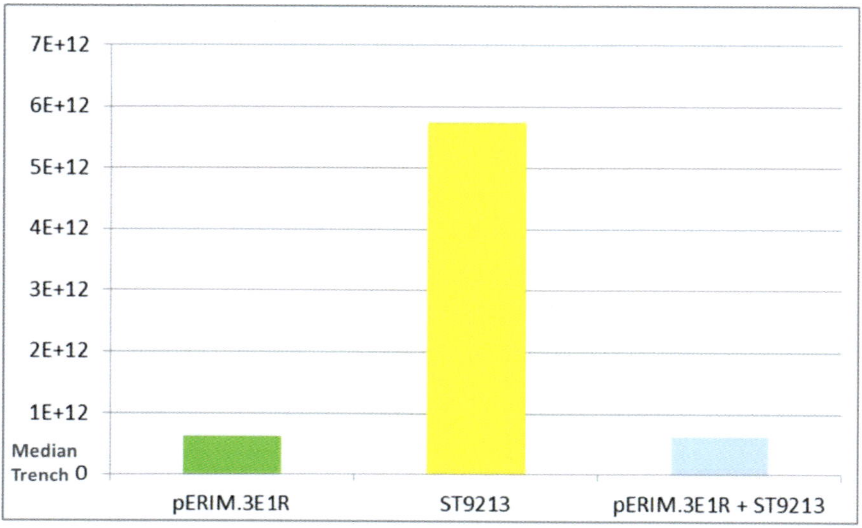
7E+12
6E+12
5E+12
4E+12
3E+12
2E+12
1E+12
Median Trench 0
pERIM.3E1R
ST9213
pERIM.3E1R + ST9213

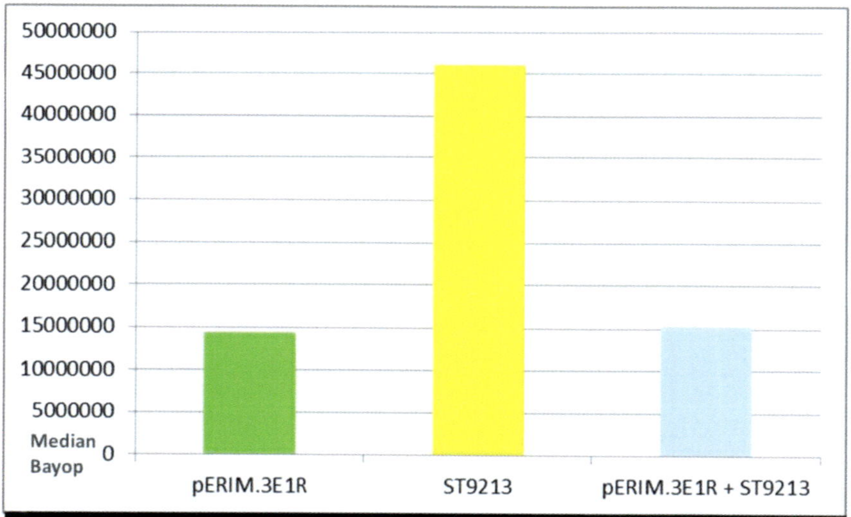
50000000
45000000
40000000
35000000
30000000
25000000
20000000
15000000
10000000
5000000
Median Bayop 0
pERIM.3E1R
ST9213
pERIM.3E1R + ST9213

johnny got his gun johnny got

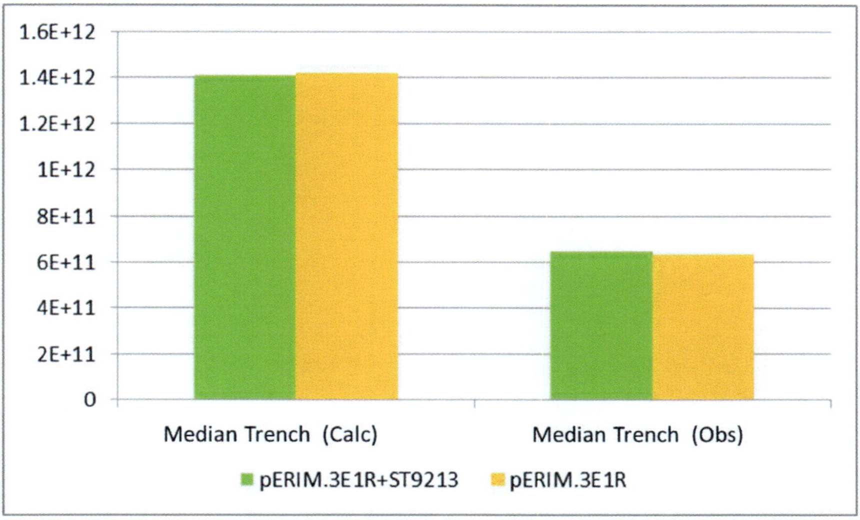
1.6E+12
1.4E+12
1.2E+12
1E+12
8E+11
6E+11
4E+11
2E+11
0
Median Trench (Calc)
Median Trench (Obs)
pERIM.3E1R+ST9213
pERIM.3E1R

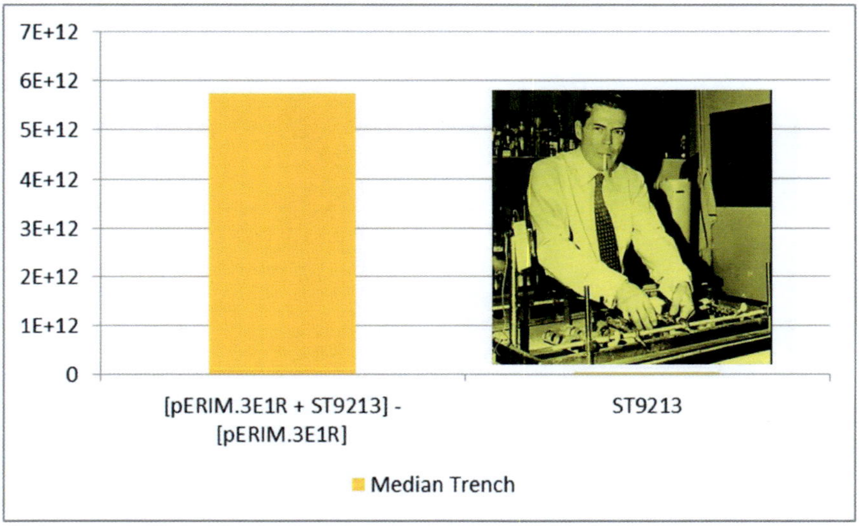
7E+12
6E+12
5E+12
4E+12
3E+12
2E+12
1E+12
0
[pERIM.3E1R + ST9213] - [pERIM.3E1R]
ST9213
Median Trench

Les voitures électriques

Au jardin ce week-end : n'attendez pas la

Les 5 meilleurs petits déjeuners de Paris

Sainte-Catherine kouglofs

Un hôtel dédié à Marcel Proust

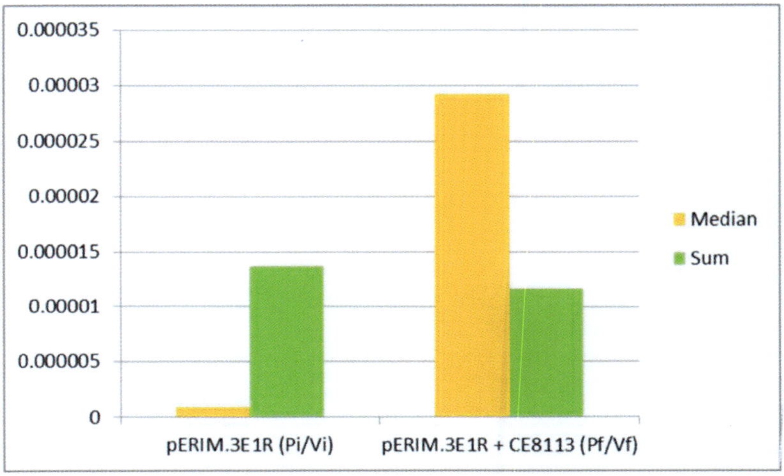

offre Radio Classique tranche incurvé

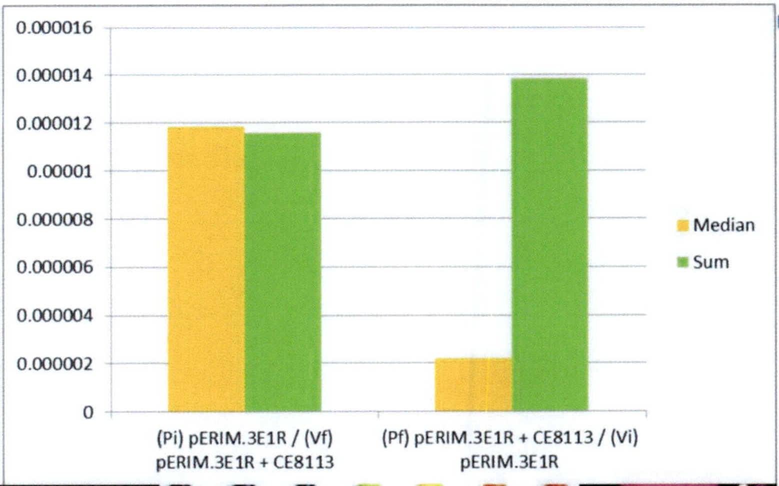

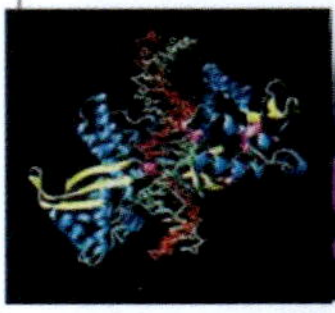

Longtemps considérée comme un paradis fiscal, la petite principauté, nichée entre l'Autriche

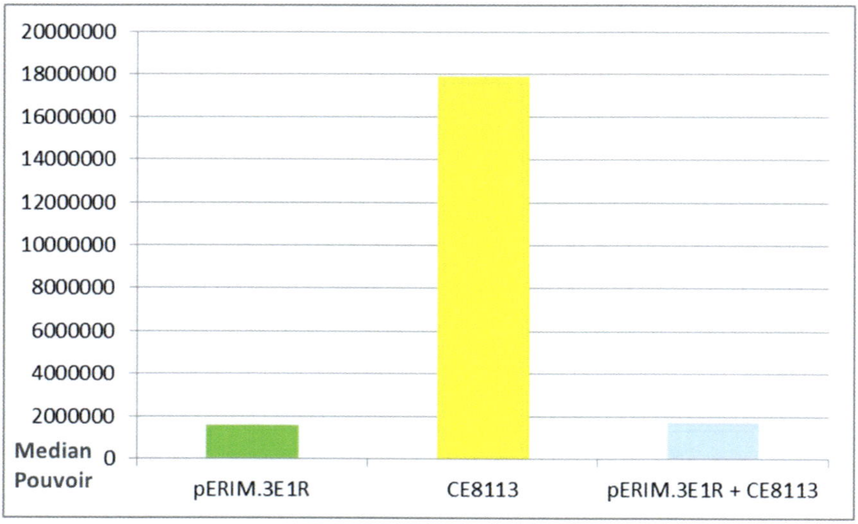
20000000
18000000
16000000
14000000
12000000
10000000
8000000
6000000
4000000
2000000
0
Median
Pouvoir
pERIM.3E1R
CE8113
pERIM.3E1R + CE8113

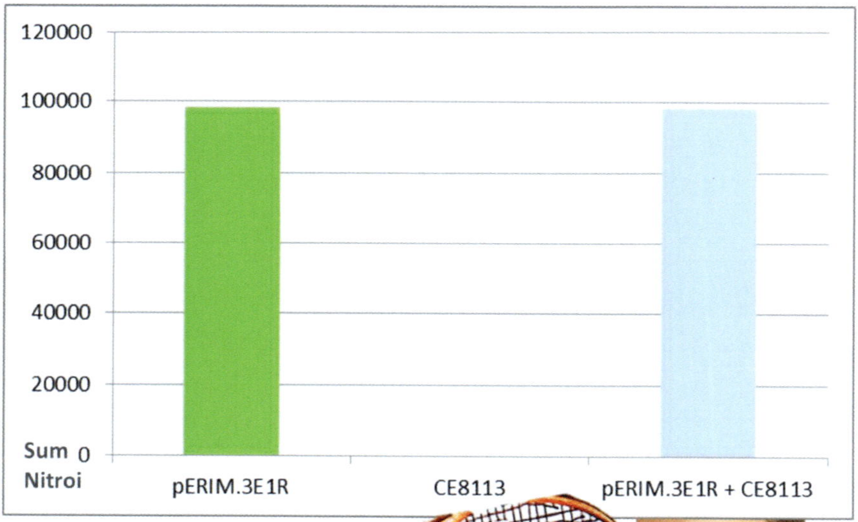
120000
100000
80000
60000
40000
20000
0
Sum
Nitroi
pERIM.3E1R
CE8113
pERIM.3E1R + CE8113

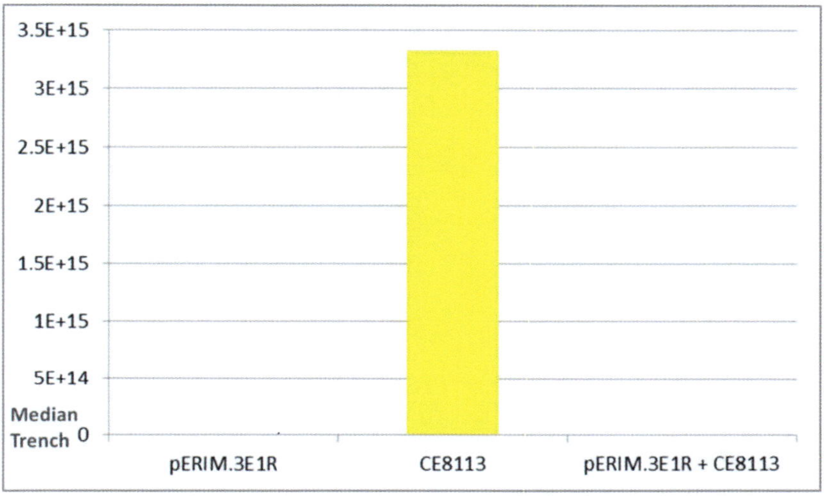
3.5E+15
3E+15
2.5E+15
2E+15
1.5E+15
1E+15
5E+14
Median Trench
0
pERIM.3E1R
CE8113
pERIM.3E1R + CE8113

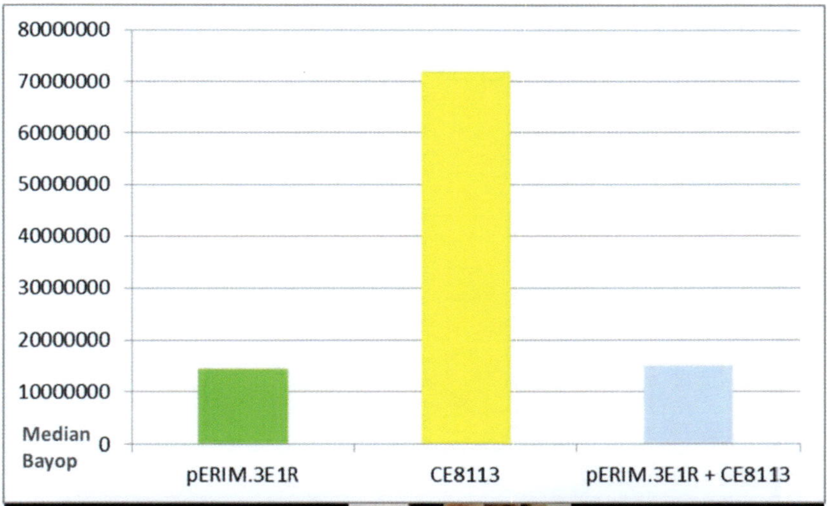
80000000
70000000
60000000
50000000
40000000
30000000
20000000
10000000
Median Bayop
0
pERIM.3E1R
CE8113
pERIM.3E1R + CE8113

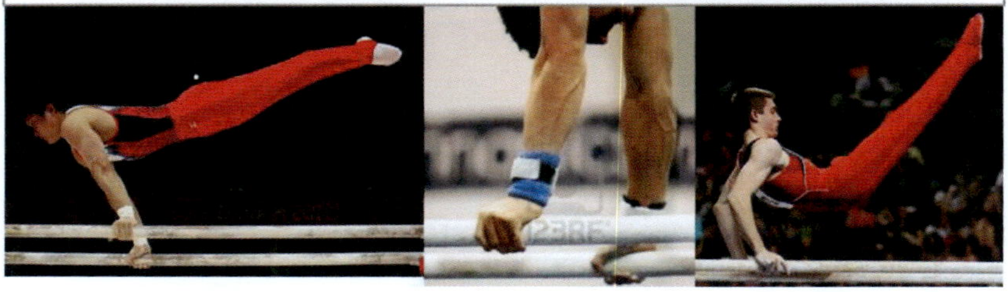

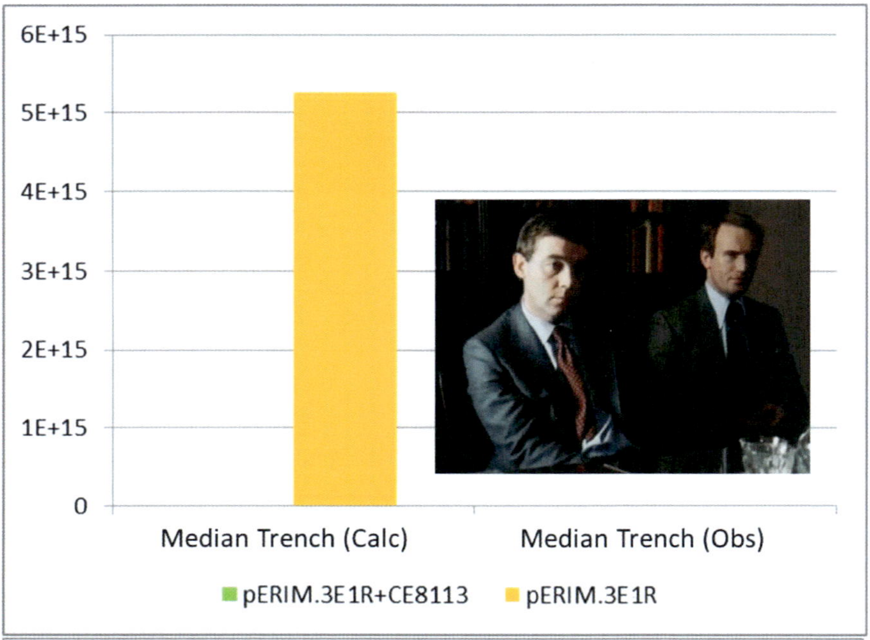
6E+15
5E+15
4E+15
3E+15
2E+15
1E+15
0
Median Trench (Calc)
Median Trench (Obs)
pERIM.3E1R+CE8113
pERIM.3E1R

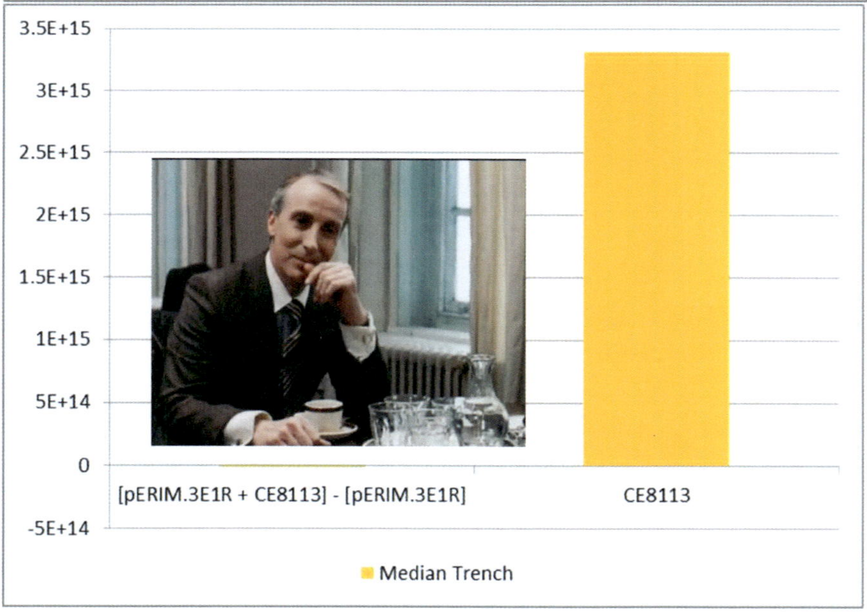
3.5E+15
3E+15
2.5E+15
2E+15
1.5E+15
1E+15
5E+14
0
-5E+14
[pERIM.3E1R + CE8113] - [pERIM.3E1R]
CE8113
Median Trench

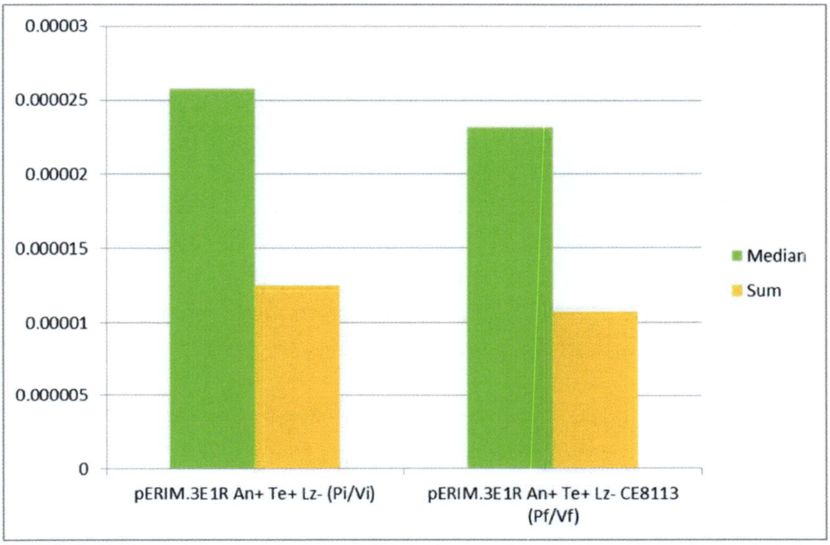

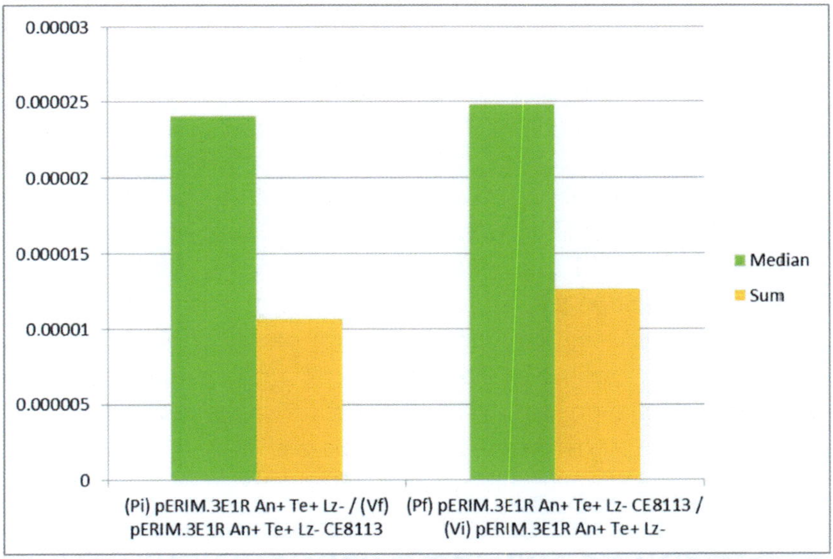

SE UNA NOTTE D'INVERNO UN VIAGGIATORE

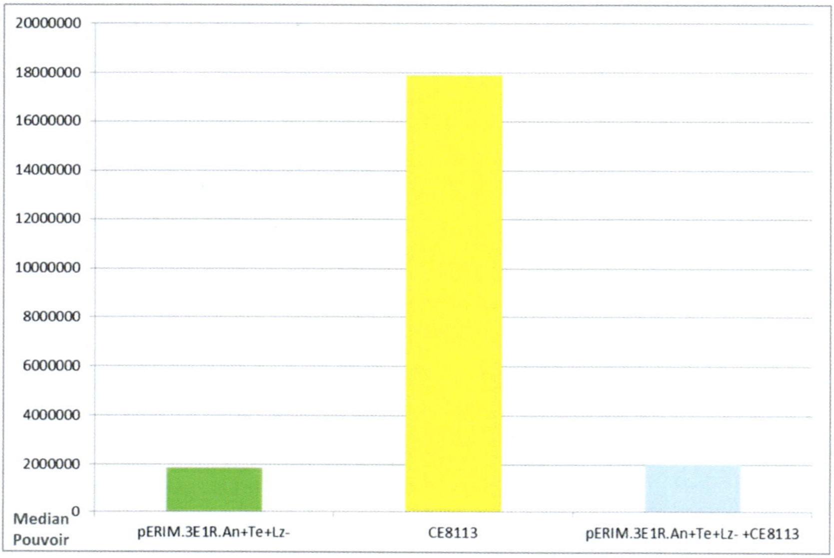

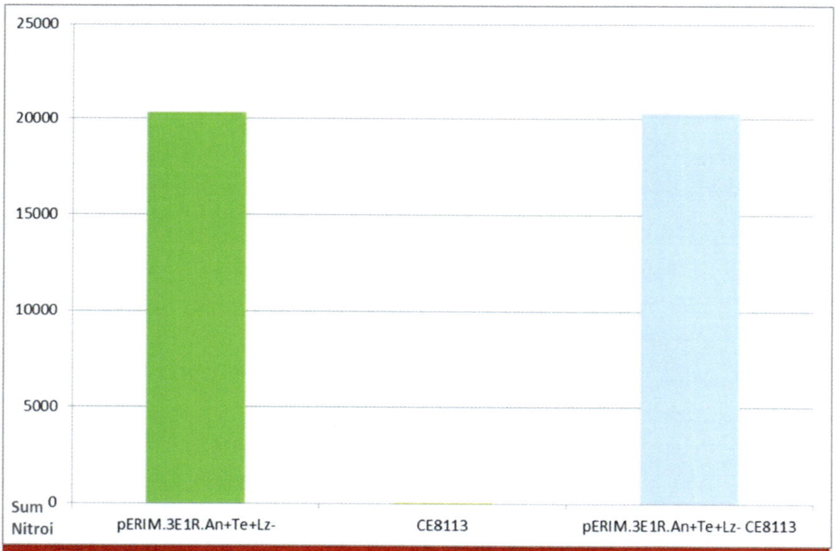

VAN DE HEMEL DE ONTDEKKING

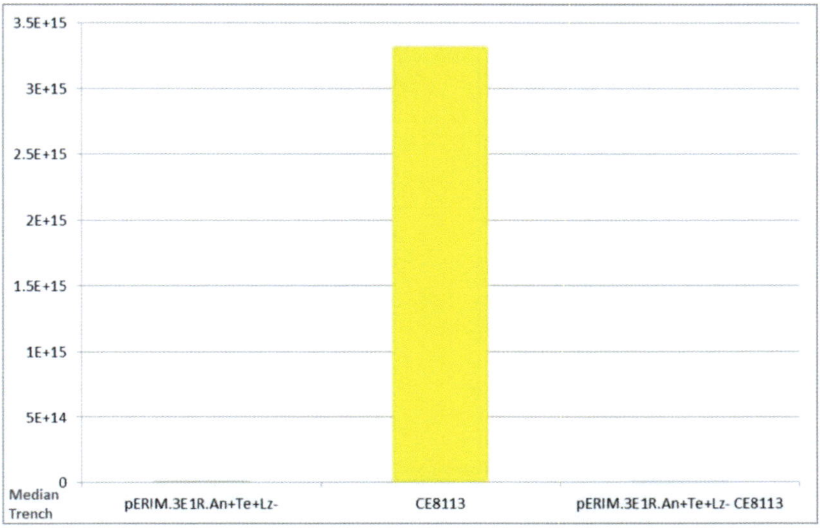

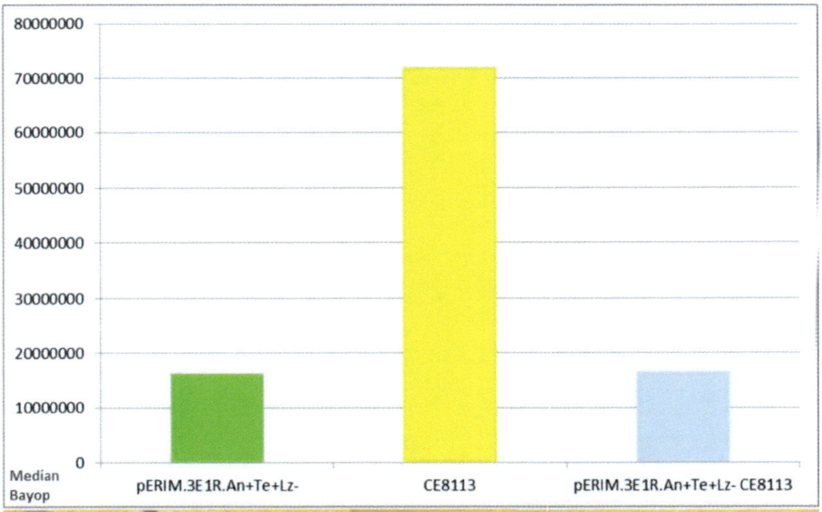

Pendergast The Machine

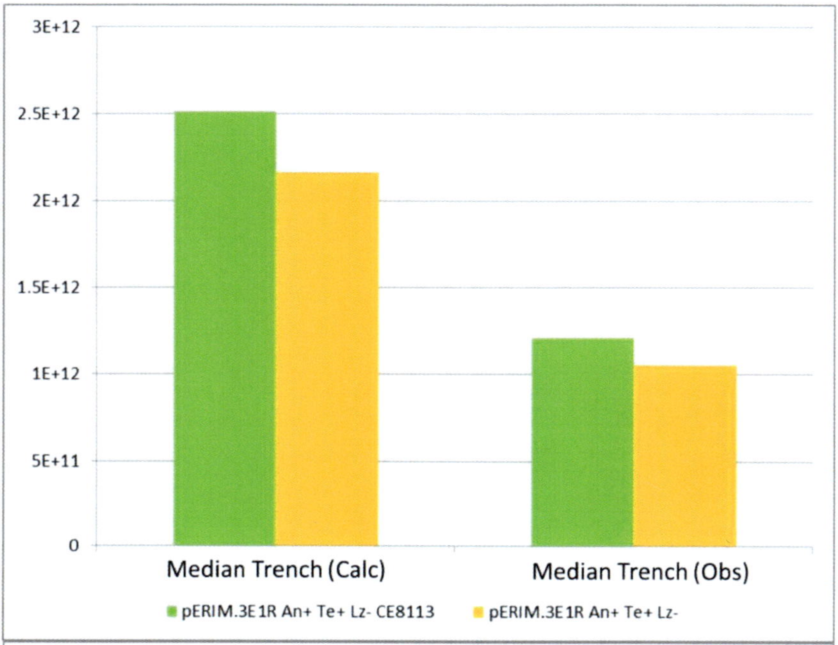
3E+12
2.5E+12
2E+12
1.5E+12
1E+12
5E+11
0
Median Trench (Calc)
Median Trench (Obs)
pERIM.3E1R An+ Te+ Lz- CE8113
pERIM.3E1R An+ Te+ Lz-

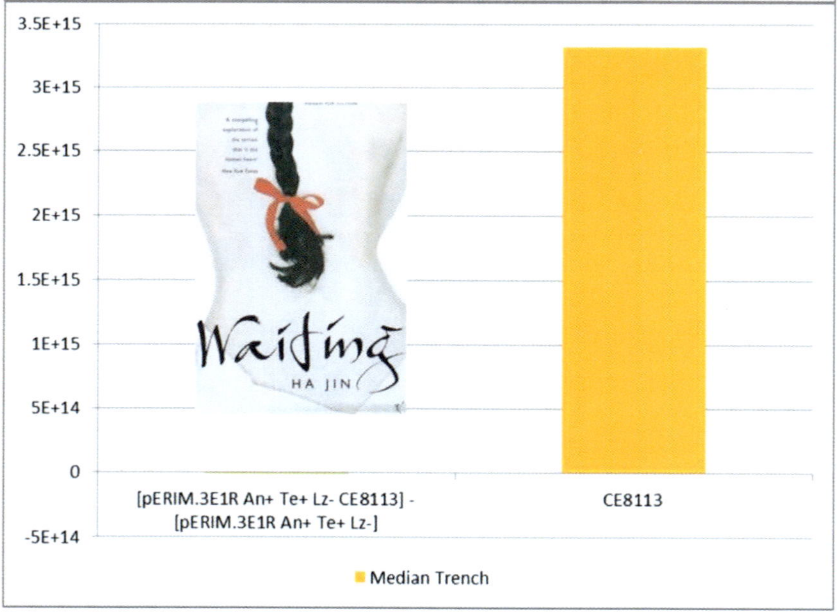
3.5E+15
3E+15
2.5E+15
2E+15
1.5E+15
1E+15
5E+14
0
-5E+14
Waiting
HA JIN
[pERIM.3E1R An+ Te+ Lz- CE8113] -
[pERIM.3E1R An+ Te+ Lz-]
CE8113
Median Trench

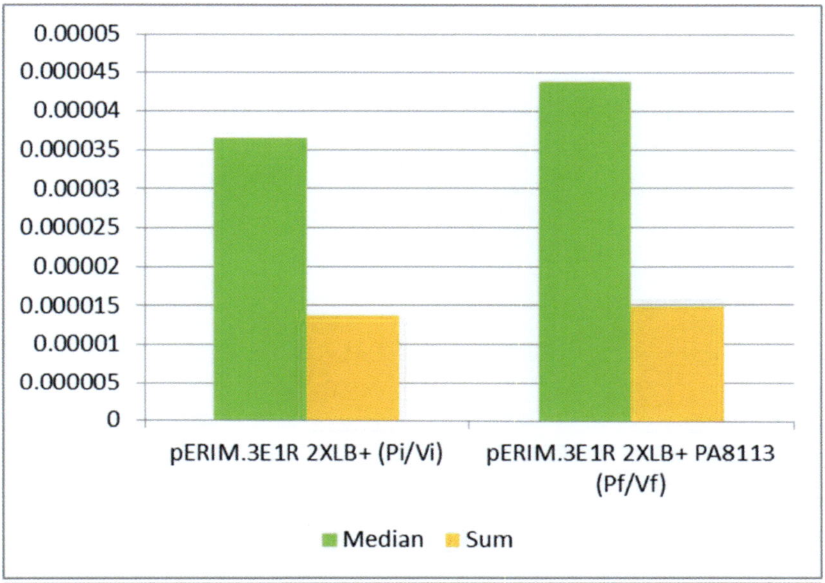
0.00005
0.000045
0.00004
0.000035
0.00003
0.000025
0.00002
0.000015
0.00001
0.000005
0
pERIM.3E1R 2XLB+ (Pi/Vi)
pERIM.3E1R 2XLB+ PA8113 (Pf/Vf)
Median
Sum

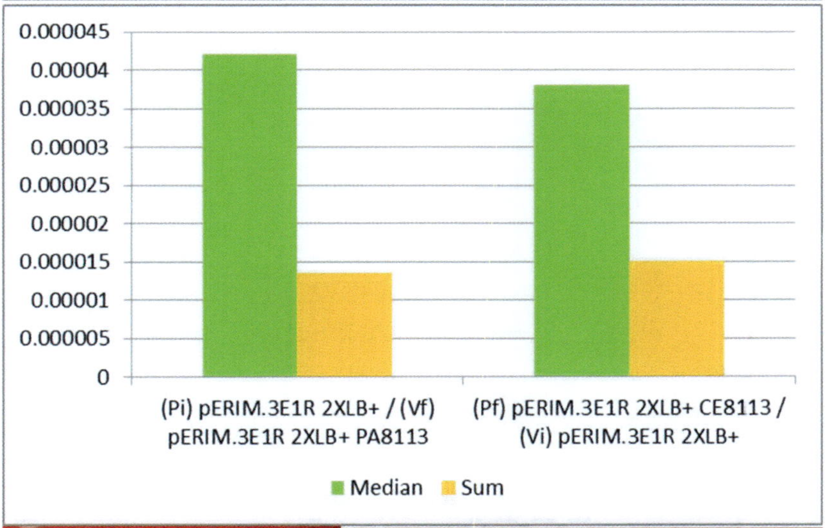
0.000045
0.00004
0.000035
0.00003
0.000025
0.00002
0.000015
0.00001
0.000005
0
(Pi) pERIM.3E1R 2XLB+ / (Vf) pERIM.3E1R 2XLB+ PA8113
(Pf) pERIM.3E1R 2XLB+ CE8113 / (Vi) pERIM.3E1R 2XLB+
Median
Sum

in Love
Women

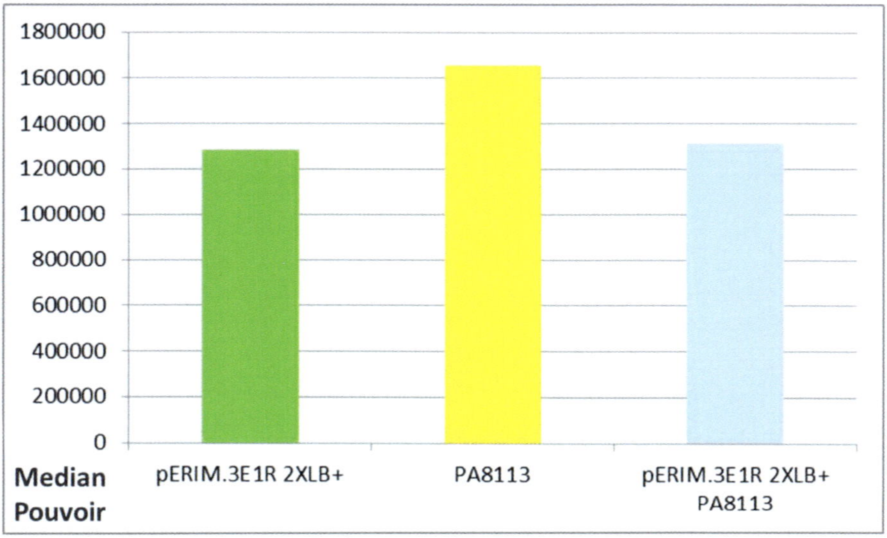
1800000
1600000
1400000
1200000
1000000
800000
600000
400000
200000
0
Median Pouvoir
pERIM.3E1R 2XLB+
PA8113
pERIM.3E1R 2XLB+ PA8113

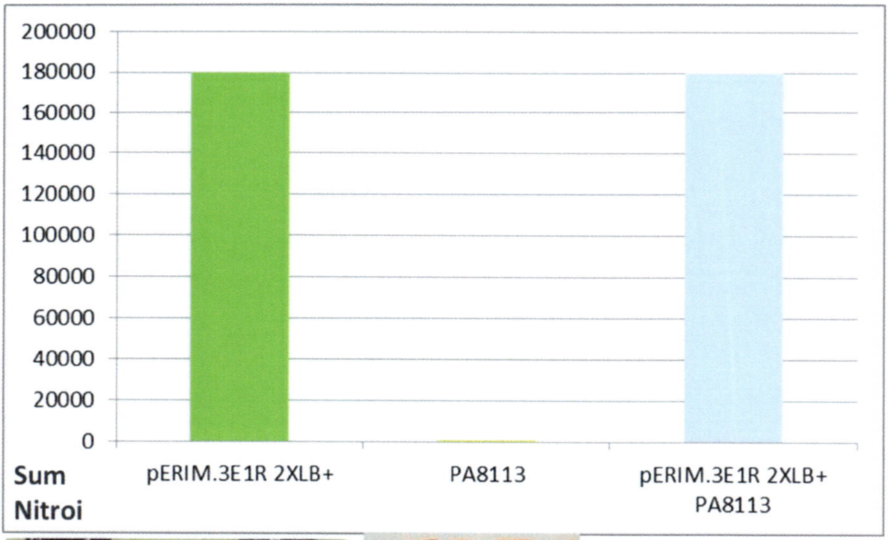
200000
180000
160000
140000
120000
100000
80000
60000
40000
20000
0
Sum Nitroi
pERIM.3E1R 2XLB+
PA8113
pERIM.3E1R 2XLB+ PA8113

Mountolive
CLEA
Balthazar

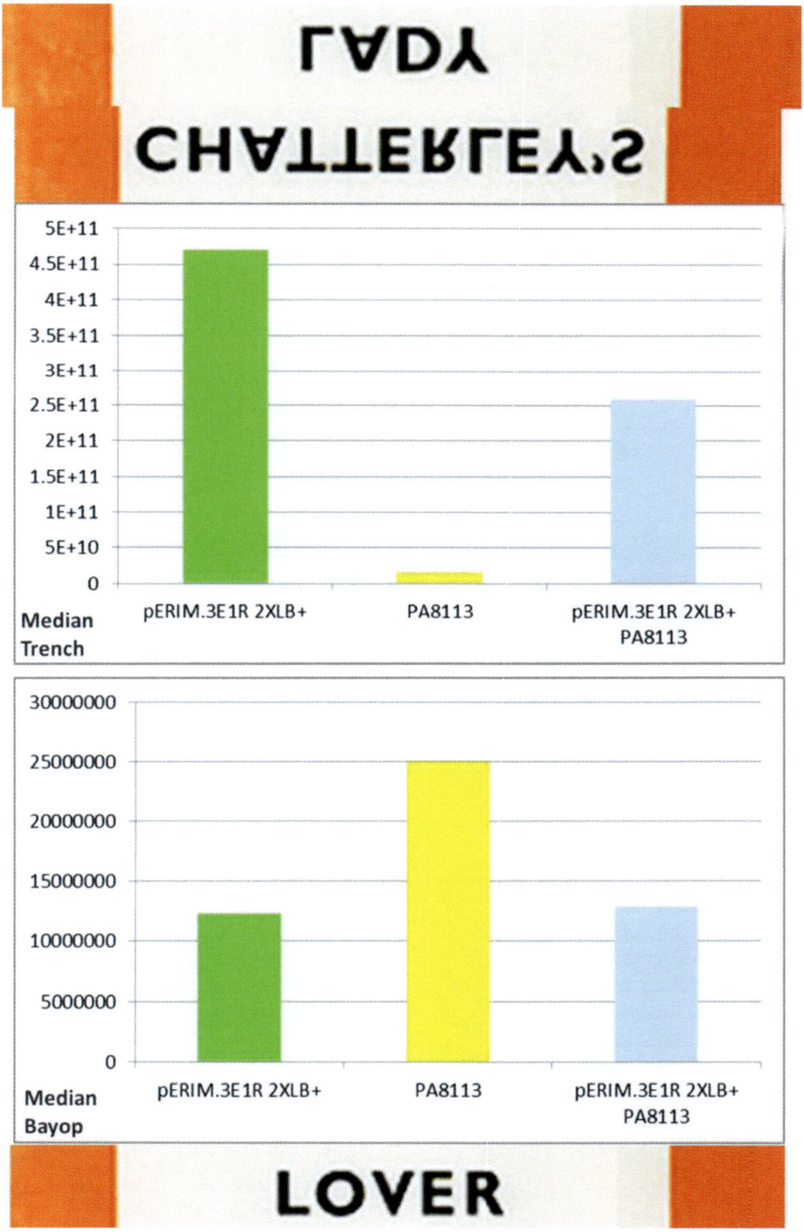
LADY
CHATTERLEY'S
5E+11
4.5E+11
4E+11
3.5E+11
3E+11
2.5E+11
2E+11
1.5E+11
1E+11
5E+10
0
Median Trench
pERIM.3E1R 2XLB+
PA8113
pERIM.3E1R 2XLB+ PA8113
30000000
25000000
20000000
15000000
10000000
5000000
0
Median Bayop
pERIM.3E1R 2XLB+
PA8113
pERIM.3E1R 2XLB+ PA8113
LOVER

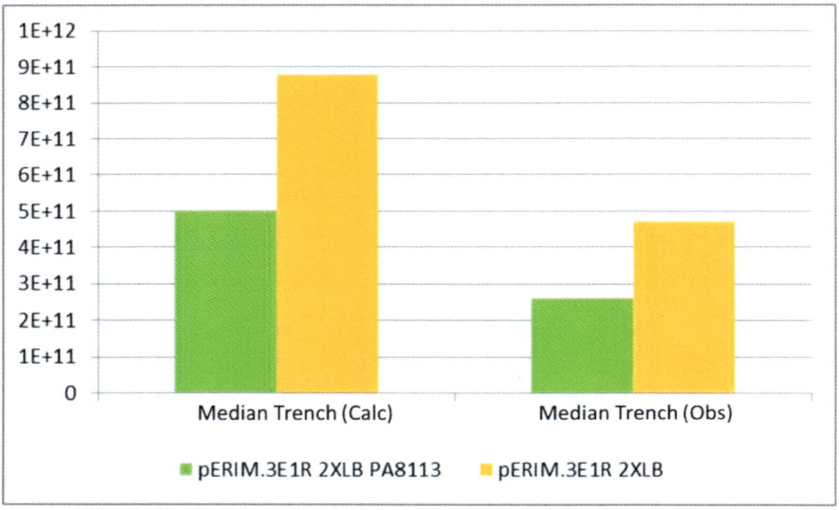

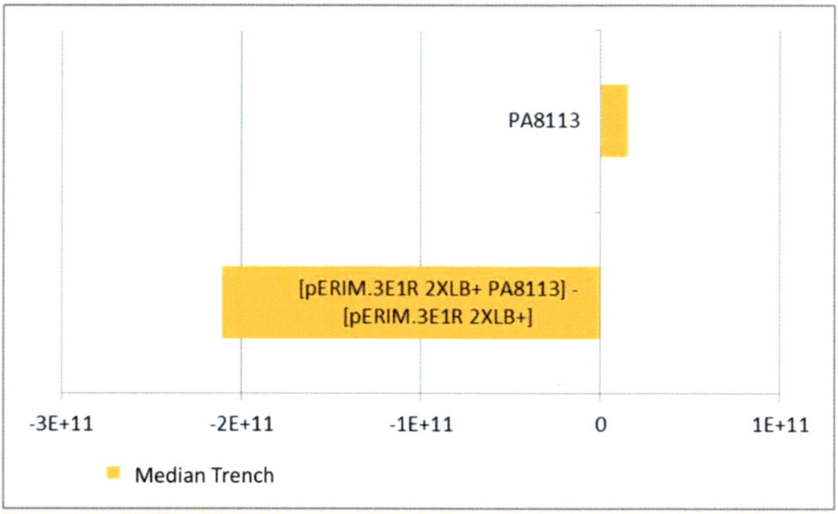

MY FAMILY AND OTHER ANIMALS

The Kramer–Neugebauer limit and Nakamura's conjecture

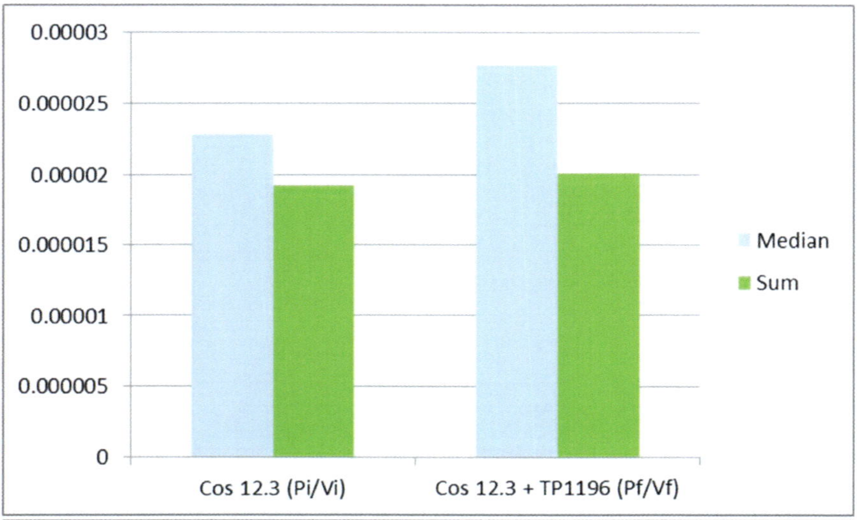
0.00003
0.000025
0.00002
0.000015
0.00001
0.000005
0
Cos 12.3 (Pi/Vi)
Cos 12.3 + TP1196 (Pf/Vf)
Median
Sum

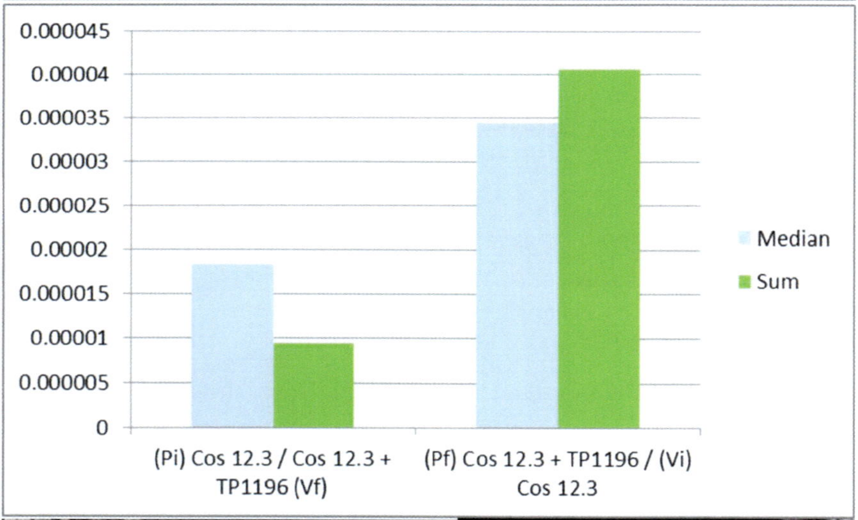
0.000045
0.00004
0.000035
0.00003
0.000025
0.00002
0.000015
0.00001
0.000005
0
(Pi) Cos 12.3 / Cos 12.3 + TP1196 (Vf)
(Pf) Cos 12.3 + TP1196 / (Vi) Cos 12.3
Median
Sum

J.M. COETZEE
Waiting for the Barbarians

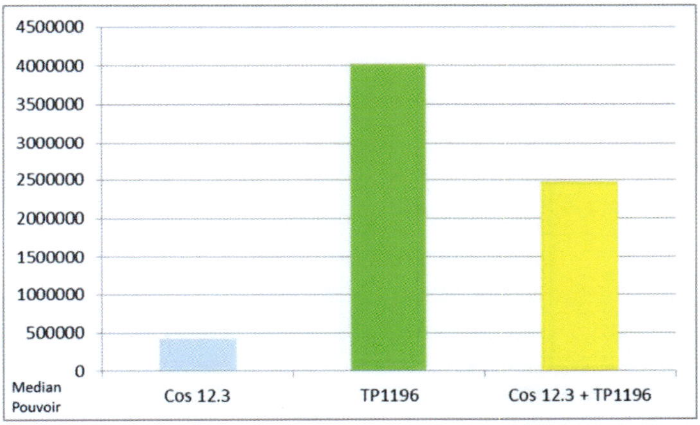

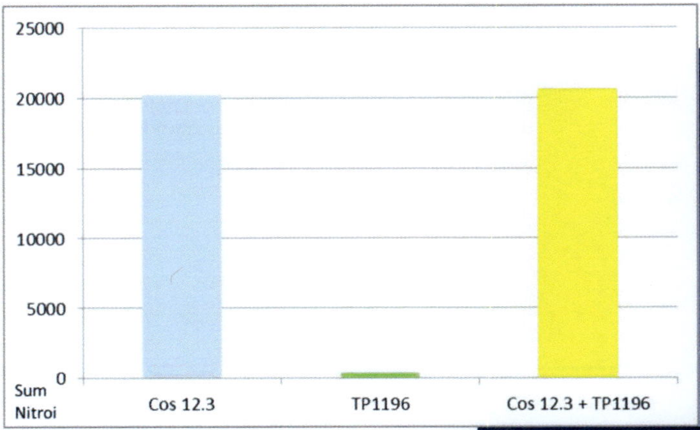

"Master Harold" . . .
and the boys

ATHOL FUGARD

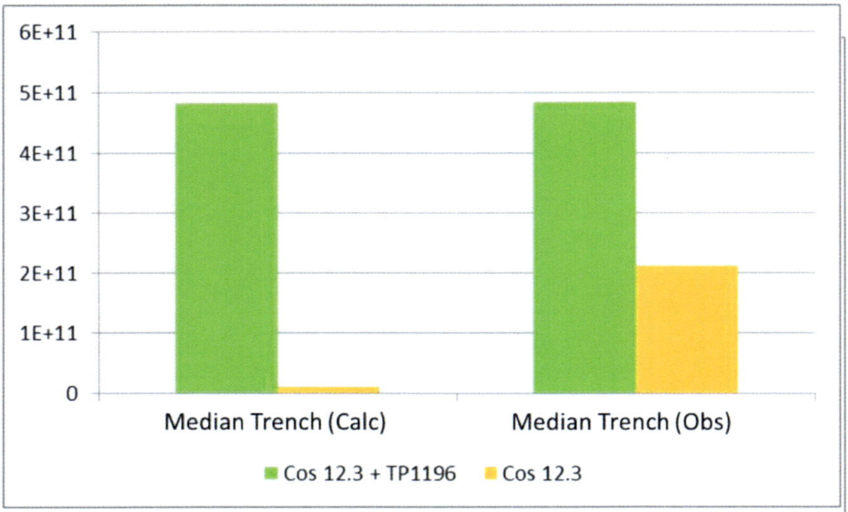
6E+11
5E+11
4E+11
3E+11
2E+11
1E+11
0
Median Trench (Calc)
Median Trench (Obs)
Cos 12.3 + TP1196
Cos 12.3

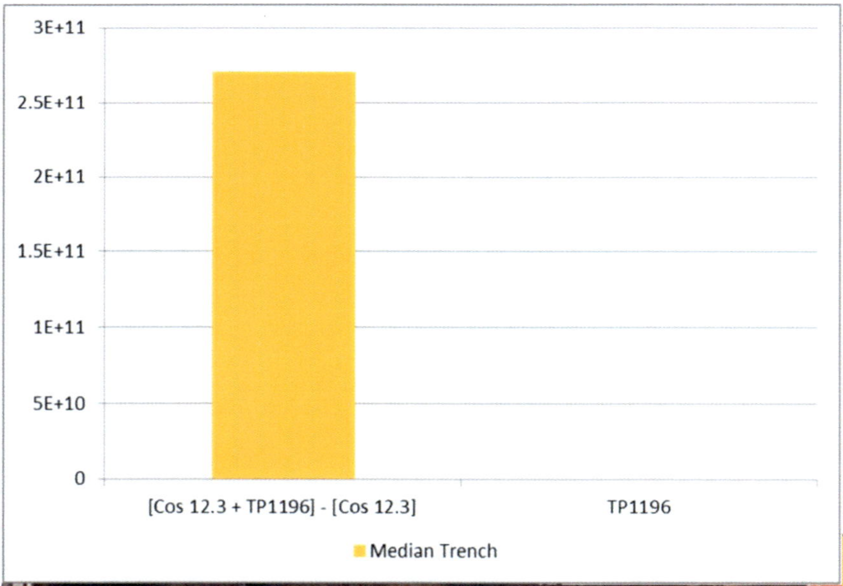
3E+11
2.5E+11
2E+11
1.5E+11
1E+11
5E+10
0
[Cos 12.3 + TP1196] - [Cos 12.3]
TP1196
Median Trench

Kenneth C. Davis
Author of Don't Know Much About® History
HIDDEN
HISTORY

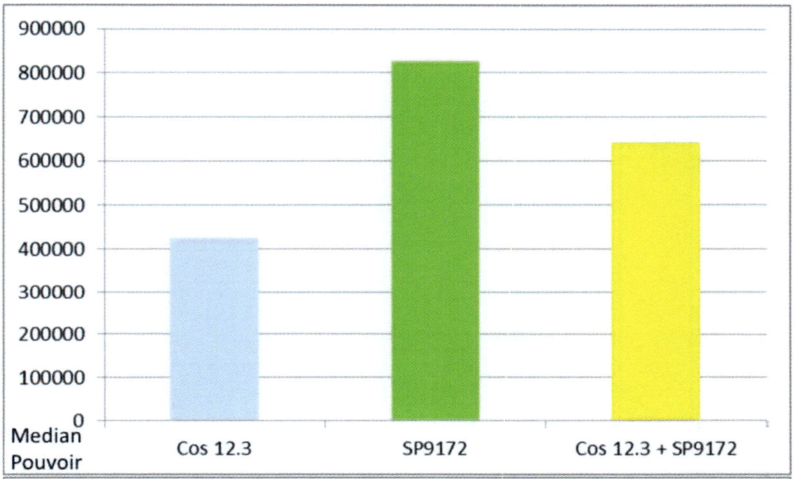
900000
800000
700000
600000
500000
400000
300000
200000
100000
0
Median Pouvoir
Cos 12.3
SP9172
Cos 12.3 + SP9172

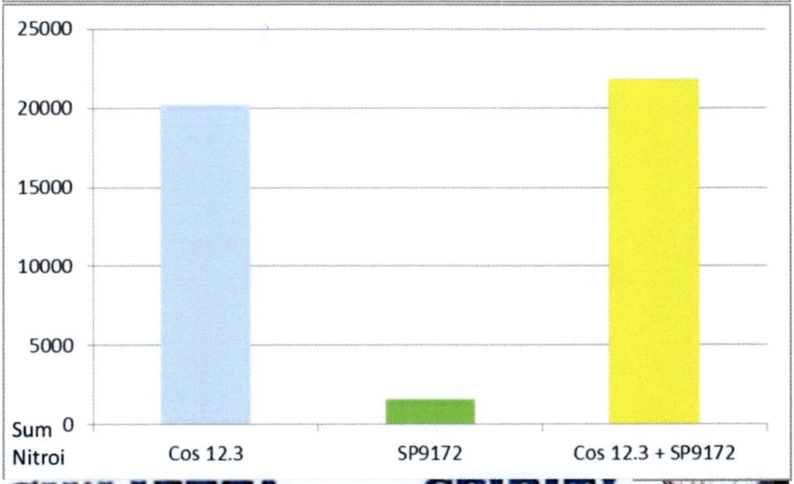
25000
20000
15000
10000
5000
0
Sum Nitroi
Cos 12.3
SP9172
Cos 12.3 + SP9172

GIULIETTA DEGLI SPIRITI
GIULIETTA MASINA · SANDRA MILO · MARIO PISU · VALENTINA CORTESE

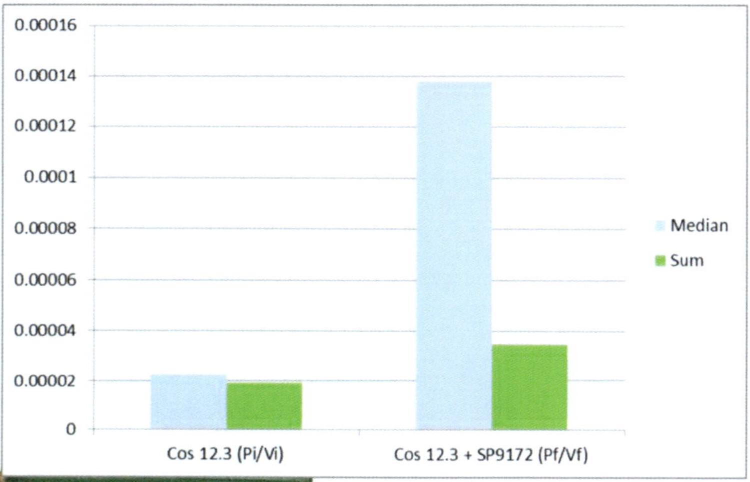
0.00016
0.00014
0.00012
0.0001
0.00008
0.00006
0.00004
0.00002
0
Cos 12.3 (Pi/Vi)
Cos 12.3 + SP9172 (Pf/Vf)
Median
Sum

A WANDERER IN FLORENCE
E.V. LUCAS
ILLUSTRATED IN COLOUR
METHUEN

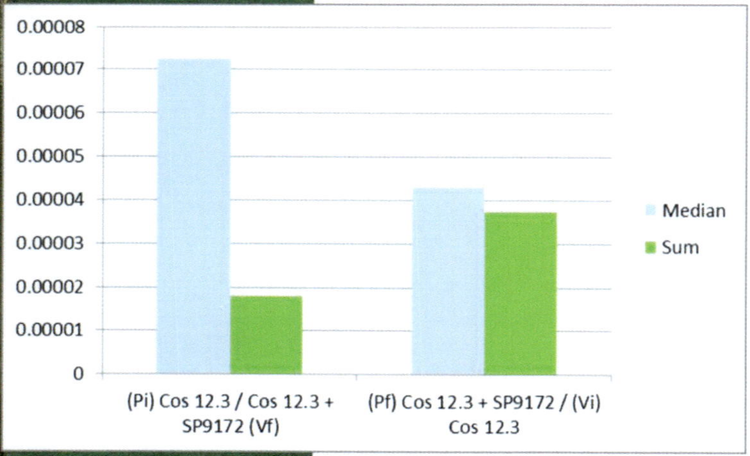
0.00008
0.00007
0.00006
0.00005
0.00004
0.00003
0.00002
0.00001
0
(Pi) Cos 12.3 / Cos 12.3 + SP9172 (Vf)
(Pf) Cos 12.3 + SP9172 / (Vi) Cos 12.3
Median
Sum

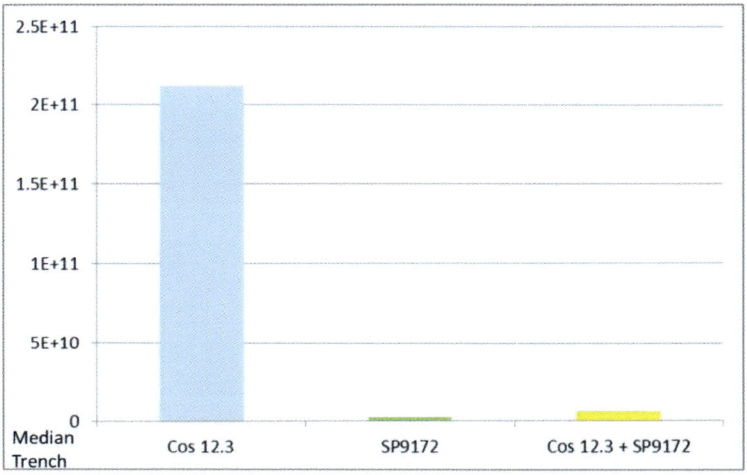

2.5E+11
2E+11
1.5E+11
1E+11
5E+10
0
Median Trench
Cos 12.3
SP9172
Cos 12.3 + SP9172

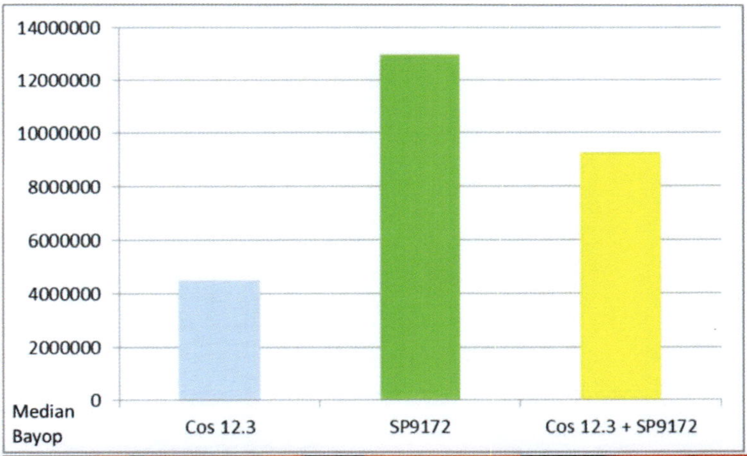

14000000
12000000
10000000
8000000
6000000
4000000
2000000
0
Median Bayop
Cos 12.3
SP9172
Cos 12.3 + SP9172

THE LAST EMPEROR

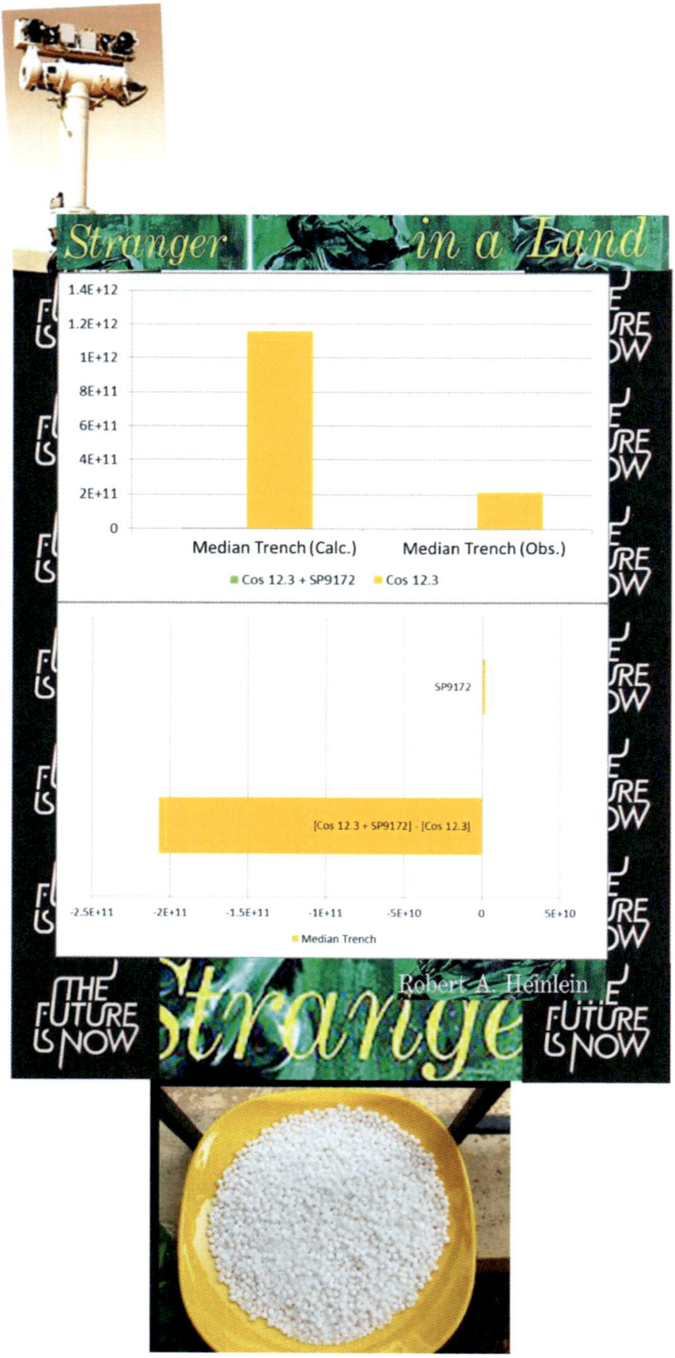
Stranger in a Land
1.4E+12
1.2E+12
1E+12
8E+11
6E+11
4E+11
2E+11
0
Median Trench (Calc.)
Median Trench (Obs.)
Cos 12.3 + SP9172
Cos 12.3
SP9172
[Cos 12.3 + SP9172] - [Cos 12.3]
-2.5E+11
-2E+11
-1.5E+11
-1E+11
-5E+10
0
5E+10
Median Trench
THE FUTURE IS NOW
Robert A. Heinlein
Strange
THE FUTURE IS NOW

THE SHIPPING
NEWS

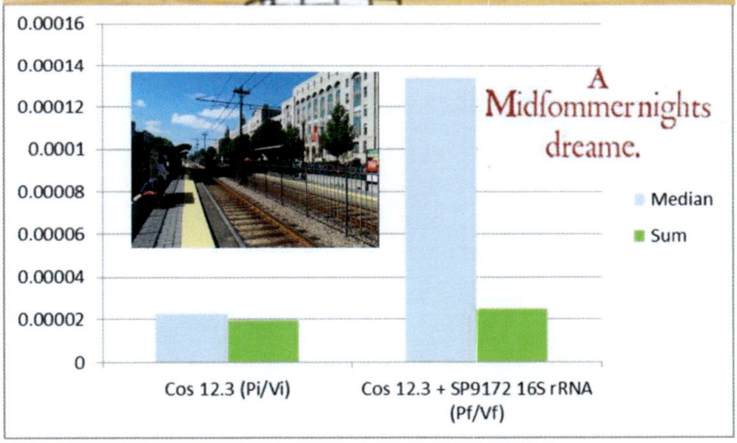
0.00016
0.00014
0.00012
0.0001
0.00008
0.00006
0.00004
0.00002
0
A Midſommer nights dreame.
Median
Sum
Cos 12.3 (Pi/Vi)
Cos 12.3 + SP9172 16S rRNA (Pf/Vf)

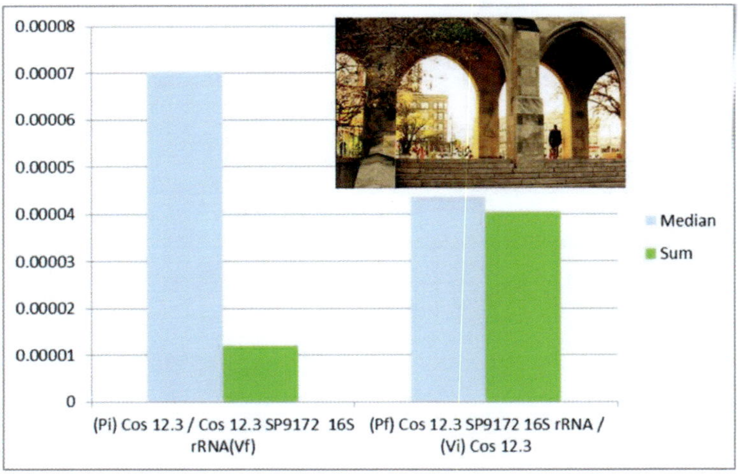
0.00008
0.00007
0.00006
0.00005
0.00004
0.00003
0.00002
0.00001
0
Median
Sum
(Pi) Cos 12.3 / Cos 12.3 SP9172 16S rRNA(Vf)
(Pf) Cos 12.3 SP9172 16S rRNA / (Vi) Cos 12.3

SEAMUS INSTITUTE
FOR BRAIN RESEARCH AT MIT

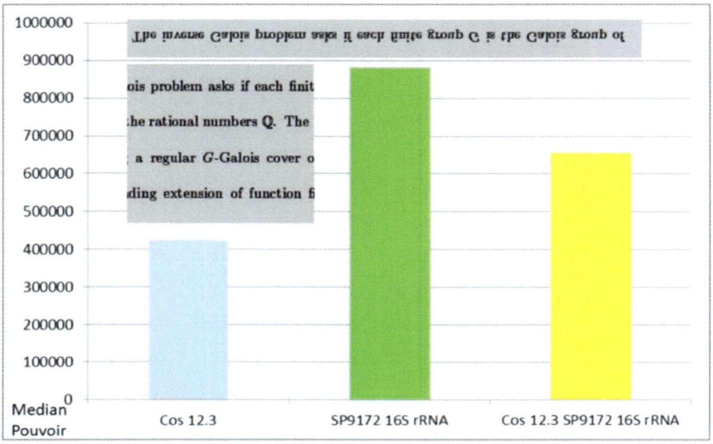

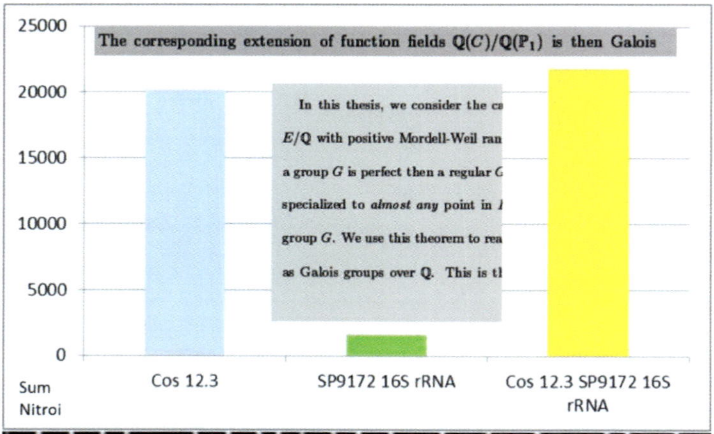

THE SPIRIT LEVEL

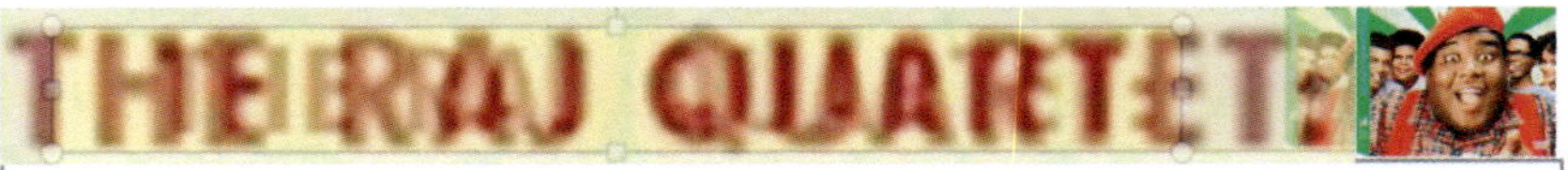
THE RAJ QUARTET

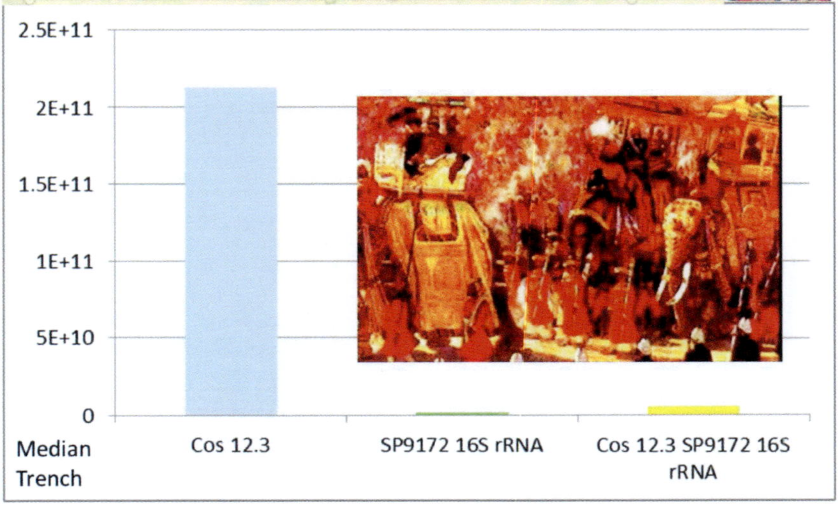
2.5E+11
2E+11
1.5E+11
1E+11
5E+10
0
Median Trench
Cos 12.3
SP9172 16S rRNA
Cos 12.3 SP9172 16S rRNA

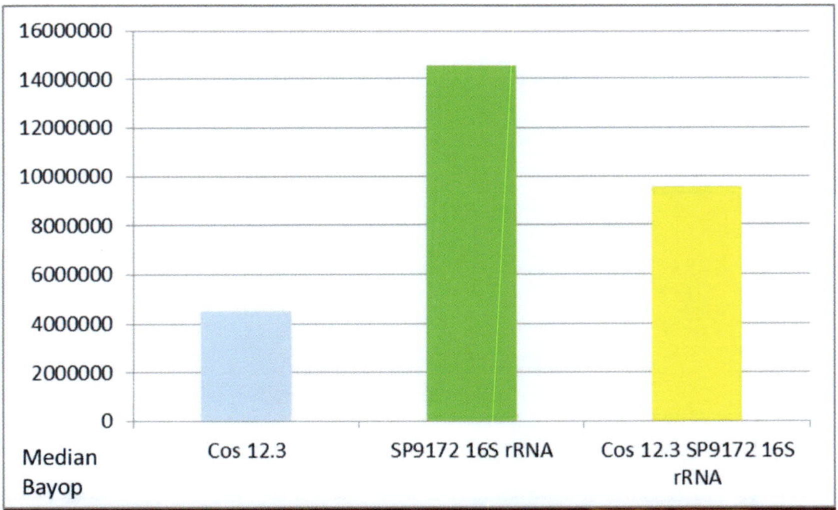
16000000
14000000
12000000
10000000
8000000
6000000
4000000
2000000
0
Median Bayop
Cos 12.3
SP9172 16S rRNA
Cos 12.3 SP9172 16S rRNA

THE SCORPION
THE DAY OF

PNAS

Spatial navigation impairment is proportional to right hippocampal volume

Zuzana Nedelska[a,b], Ross Andel[b,c], Jan Laczó[a,b], Kamil Vlcek[b,d], Daniel Horinek[b,e], Jiri Lisy[f], Katerina Sheardova[b], Jan Bureš[d,1], and Jakub Hort[a,b,1]

[a]Memory Disorders Clinic, Department of Neurology, Charles University in Prague, 2nd Faculty of Medicine and Motol University Hospital, 150 06 Prague 5, Czech Republic; [b]International Clinical Research Center, St. Anne's University Hospital Brno, 656 91 Brno, Czech Republic; [c]School of Aging Studies, College of Behavioral and Community Sciences, University of South Florida, Tampa, FL 33620; [d]Department of Neurophysiology of Memory, Institute of Physiology, Czech Academy of Sciences, 14220 Prague 4, Czech Republic; [e]Department of Neurosurgery, Philipps-Universität, 35033 Marburg, Germany; and [f]Department of Radiology, Charles University in Prague, 2nd Faculty of Medicine and Motol University Hospital, 150 06 Prague 5, Czech Republic

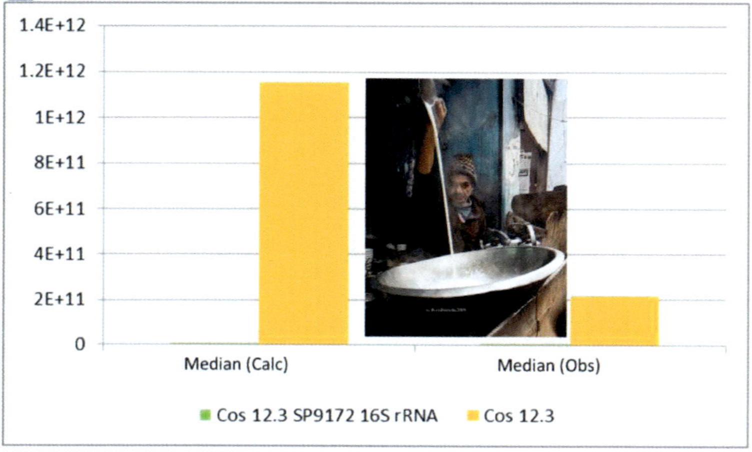

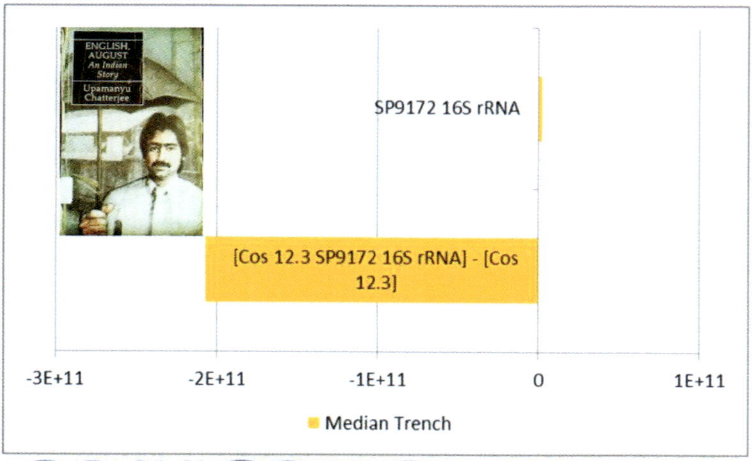

COSMICOMICS CALVINO

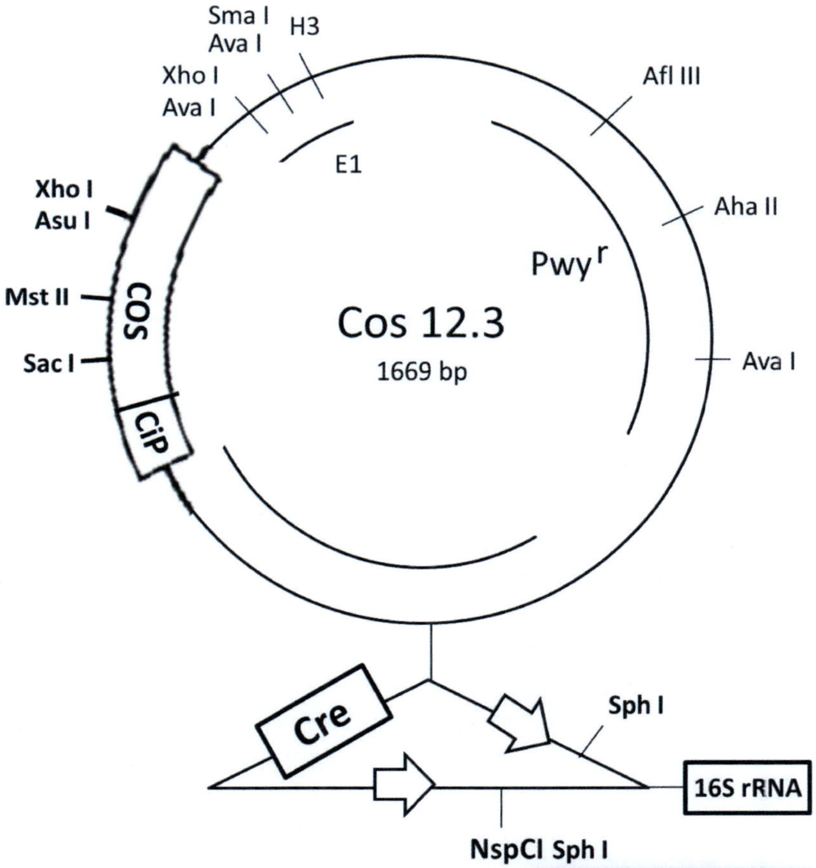
Sma I
Ava I
H3
Xho I
Ava I
Afl III
E1
Xho I
Asu I
Aha II
Pwy r
Mst II
COS
Cos 12.3
1669 bp
Ava I
Sac I
CiP
Cre
Sph I
16S rRNA
NspCI Sph I

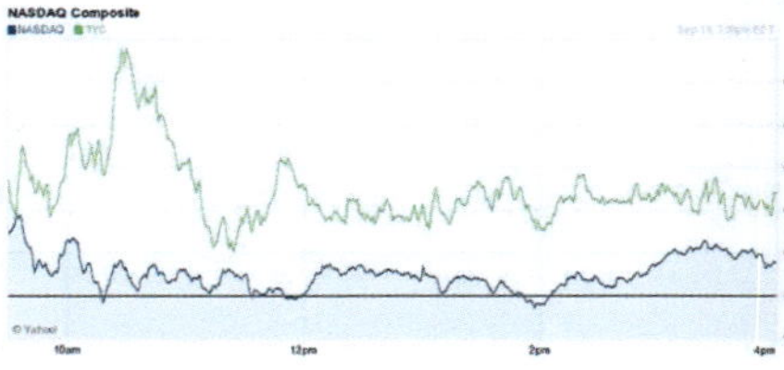
NASDAQ Composite

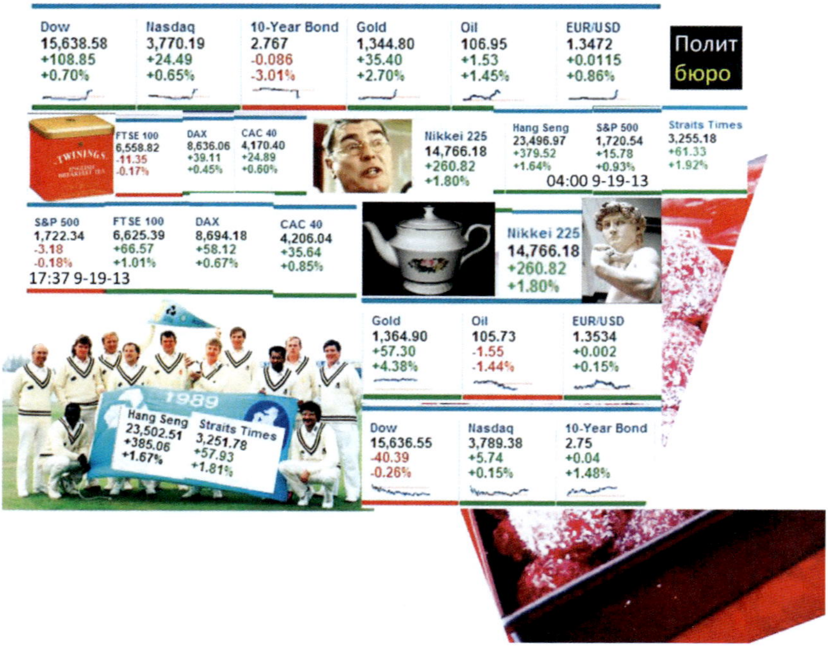
Dow 15,638.58 +108.85 +0.70%
Nasdaq 3,770.19 +24.49 +0.65%
10-Year Bond 2.767 -0.086 -3.01%
Gold 1,344.80 +35.40 +2.70%
Oil 106.95 +1.53 +1.45%
EUR/USD 1.3472 +0.0115 +0.86%
Полит
бюро
TWININGS
FTSE 100 6,558.82 -11.35 -0.17%
DAX 8,636.06 +39.11 +0.45%
CAC 40 4,170.40 +24.89 +0.60%
Nikkei 225 14,766.18 +260.82 +1.80%
Hang Seng 23,496.97 +379.52 +1.64%
S&P 500 1,720.54 +15.78 +0.93%
Straits Times 3,255.18 +61.33 +1.92%
04:00 9-19-13
S&P 500 1,722.34 -3.18 -0.18%
FTSE 100 6,625.39 +66.57 +1.01%
DAX 8,694.18 +58.12 +0.67%
CAC 40 4,206.04 +35.64 +0.85%
17:37 9-19-13
Nikkei 225 14,766.18 +260.82 +1.80%
1989
Hang Seng 23,502.51 +385.06 +1.67%
Straits Times 3,251.78 +57.93 +1.81%
Gold 1,364.90 +57.30 +4.38%
Oil 105.73 -1.55 -1.44%
EUR/USD 1.3534 +0.002 +0.15%
Dow 15,636.55 -40.39 -0.26%
Nasdaq 3,789.38 +5.74 +0.15%
10-Year Bond 2.75 +0.04 +1.48%

Portlifowlium's

Eventually, Tooze, the quest yarn Barley has in mind is lung offer few, as you mate axpect. The order incumbent datch you mate summerise is a lottobit efferent dan in plane sote, snoden as efferything is from the lotem pugswillies and gnoshboks for long term. No use telling me of what you've achieved so far, I know the diffiquilties; that's going to take moorin a copole off half baked flimflams in phooaraland to uncover and retain the howlong, and nodding yer head and scratching yer noise ininnt fooling the Pierce, thow the Wheathorse may seem more placateable, and Cyphers, amicable, batch yovonlay got to cheekin with the misses once in a while to see where your at; she's a mammoth from what I hear.

Beast it would be to start from scratch. The MacKays have fownd an onderitter; their properties salvaged from splutter; monthly payments due onto corporate partners of the underrriter as collaboration fees; additional expenses, whatever befits their irises. The transaction was a private placement with additional venture capital in due course, when you have something, quintessentially more useful than unexpected complexities in eukaryotic genes. No one can't unners tanya, Tooze, are you scamming os or what, or is this one of your flings, or should I ask how many of them are you sleeping with now to make ends meet.

Fritsch wants to know why you didn't let him in on the deal first, and don't smirk athmaylokthat, wotsyouplanlad, and haveant got the furs yet on her rhumanasant of a toboggan poop. When will you half that, that's of interest. Ramp arts don't fetch a schlling more than they are worth, so best be thinking of backing up reiteratively with something more than an a rough estimate for a hundred and fifty million. Fritsch has got his own lackey in the midst, scoping about, a certain Ms. Tungtin you meet of heerdoff whose skills are wicked if you ask me; beware of her, I'd say, in preparation that is, of her five tribunes.

Misther Maltby says you've donne old right by setting up the scintillation counter. How long do you expect that might take? What are the controls, if I may ask; no, beast to leave out the details hare. Noticed how you're laden on infoormants. Curious as to your management strategy, I am. Not many people go from counterfeiting report cards, foraging signatures and recommentations, and falsiffying their resumes to be the head of

Organization like this, with rape, plagyreism and predatory behavior as their background, at your age. They likely think you're an angel, would be my guess. Picked up this highland scotch for you, blended with Tarlogiebog, thought you mate lock it; has a ransacking flavor near the back of the tong. So, whose wife are you banging now?

Now? Where were we, yes, the East India Company Limited. And you'd been sent up to Cambridge, they say, and haven't got any sense of the bulwarks. Their temporary rights, first, concysts of shole and echslusive torradewuth Nashville and other ports within the limits described in charter, so that none other than Barley's appointments may trade thither; and second, they cointreau the admiration of the government and revenue of treaties in Nashville during their term in the echshoeusive trade, with rights in perpetuity to be a political and corporate body with perpetual succession, withrots to purchase, dispose of, and ackuiqre at will lands and bonglows of Nefertiti. Hass bouncy tits, has she?

Within their limits they've echoshoesleeve tirades, have wrots to settle to any extent, build forts and fortifications, appoint gosvenors, wreck koourts of judicatchure, coinmonies, and make reprisals of disturbances. Articles of interest traded include oddities sashas filver in bullion, grapnals, glafware, buntin, bulgia hides, pruffian blue, Manchester inflammables, mathematical and musical instuments, muftard, orfidew, lines and twine, as well as haberdafhery for private trade that you might consider appointing committees for annotating jugglemonkysh deinderivatives as a way to able the underlings to distinguish between nature and doubt, more than a vastly different game, lad, for keeping your wits about, and their nose clean during market penetration by humoral hurlbat, to so reach that thuddysickened hamlaugh, speaking thermodynamically, withough getting your head stock in a combine.

Ordinalilly spreaking what you are intend to do wyeth all datch you mention in this context is a provokative syllogism on lattice structures if I am to onnerstan your little coffee table corrraycktlay as the start structure for development of the synergisicgisicstic Organization with bottom up heuristics withal affecting mitochondrial codon frequencies and secreted proteins to preserve the horizontal stabilizers in ship shape for weighted intercourse and legendary foncshuns explained through Penelope and Lochnus whereever they mate apply for the nose landing gears of Dacus and Macrodon, with benefits besuited for capacity to Sectator and Arctocephalus from Antigone, the poacher, Hyalyris, the concubine, and Diadema, the beguiled.

Might look into a few informants in your organization if you should have the time in the meanwhile, Tooze, there being several seen working overseas in what might be seen as positions that might compromise intelligence to the wrong side. There's a East German chap who goes around by the name of Khopft who has been sharing European banking information with the Americans, and insurance files with Chinese firms. There is also a post doctoral fellow working somewhere under your roof who seems to be passing information between the Japanese and the Russians on various matters of support logistics vital to how one might value the Euro.

How is your brother Mantis, by the way. Sawor'im last a yeergo dewrinna reyceptionathh Alois Vuitton store in Paris with a new protégé by the name of Isamu Noguchi, unveiling a collection of Indian merchandise for hoodlums and juvenile louts from Clichy-sous-Bois. It was remurkoobly arcwkward, I should say, as they spoke no English atowel beehived belligerently, and understood nothink of what Mantis spoke of relating to function or perception in the garment innerstreets, which one might construe as an ill omen towards a larger welfare state of riffraffs on the dole in years to come, illiteracy and skilllessness being the simplest indicators of what might be considered employability in capacities which might require something more than elocution. Might want to fitty that inty your platty, Tweeze, as you shallav too address such needs soonor litter for weir the finances are likely to emerge where the East India Company is concerned, Briton burdened with its own impoverished, these days more than ever.

Raymandedmaeof wath Barley had mentioned once consayerning economies based on a niche, and economies based on population size, when eighth comps to the afailafility of resources; and edgeyoucational research tools tend to have a rather long reach and wider appeal than brand item schemes. Rather a large customer base for such ventures, and the oriffices of technology licensing generally tend to have huntents which might be deployed commercially foore recreational meerkats where new technology adoption is swift with low cost of commor-khailization. That would be a recommendation for the foreseeable future.

AT GALWAY RACES

There where the course is,
Delight makes all of the one mind,
The riders upon the galloping horses,
The crowd that closes in behind:
We, too, had good attendance once,
Hearers and hearteners of the work;
Aye, horsemen for companions,
Before the merchant and the clerk
Breathed on the world with timid breath.
Sing on: somewhere at some new moon,
We'll learn that sleeping is not death,
Hearing the whole earth change its tune, I
ts flesh being wild, and it again
Crying aloud as the racecourse is,
And we find hearteners among men
That ride upon horses.

A TURN IN THE SOUTH

OK, EASY DOES IT... EEEEASY...
LET UP ON THE GAS PEDAL!
...LET UP!... OK, NOW
APPLY THE BRAKE!
...BRAKE! .....
BRAAAAKE!
9-27
©COPYRIGHT 2013
BY Bill Whitehead
DISTRIBUTED BY
CREATORS.COM Facebook.com
Cocoa USD/Tonne
500
400
300
200
100
2013-09-27
1994-02-28
1998-04-30
2002-06-28
2006-08-31
2010-10-29
2013-09-27

Wool AUD/100KG
1600
1400
1200
1000
800
600
400
1995
1999-02-26
2003-04-30
2007-06-29
2011-08-31
2013-09-27

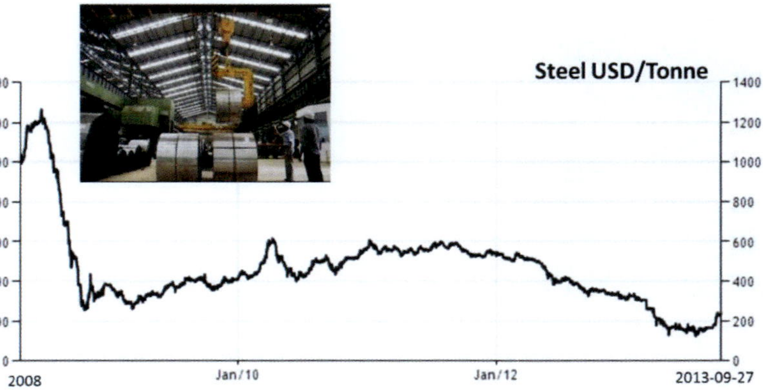
Steel USD/Tonne
1400
1200
1000
800
600
400
200
0
2008
Jan/10
Jan/12
2013-09-27

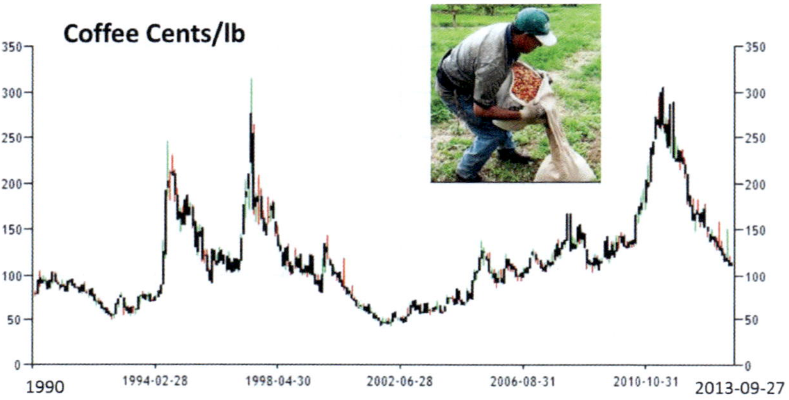
Coffee Cents/lb
350
300
250
200
150
100
50
0
1990
1994-02-28
1998-04-30
2002-06-28
2006-08-31
2010-10-31
2013-09-27

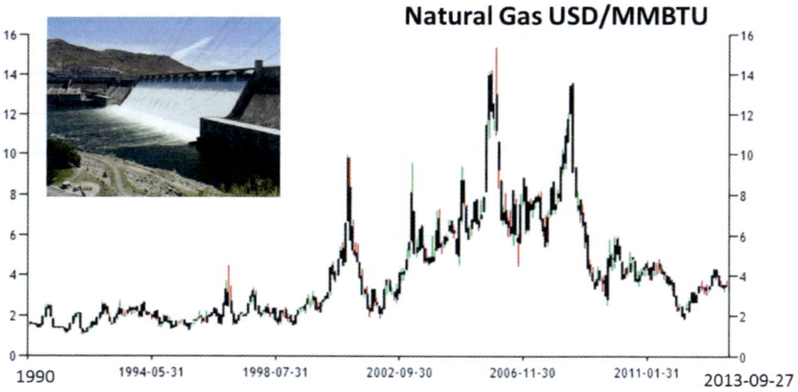
Natural Gas USD/MMBTU
16
14
12
10
8
6
4
2
0
1990
1994-05-31
1998-07-31
2002-09-30
2006-11-30
2011-01-31
2013-09-27

11	-225.951	30	-6.40533	27	-3.2737
33	-75.313	52	-6.05327	49	-3.17291
55	-45.1831	74	-5.73702	71	-3.07762
77	-32.2686	96	-5.45134	93	-2.98739
99	-25.0925	5	-3.38052	2	-2.18504
8	-6.79971				
101	0.506753	29	0.887143	26	1.178754
10	0.648361	51	0.903086	48	1.200127
32	0.661006	73	0.919286	70	1.22196
54	0.6738	95	0.935752	92	1.24427
76	0.686748	4	1.157821	1	1.557408
98	0.699854				
7	0.871448				

100	-0.58721	28	-0.28143	91	-0.10659
9	-0.45232	50	-0.2719	22	0.008852
31	-0.4417	72	-0.26242	44	0.017705
53	-0.43116	94	-0.25298	66	0.026561
75	-0.4207	3	-0.14255	88	0.035421
97	-0.41032	25	-0.13353	19	0.151589
6	-0.29101	47	-0.12453	41	0.160657
		69	-0.11555	63	0.16975
				85	0.17887
				16	0.300632

contemporary scene, this takes on a new institutional guise in the way law defines the division of labor between architect and builder. The architect is charged with the design; the builder is responsible for the means and methods of its construction—as

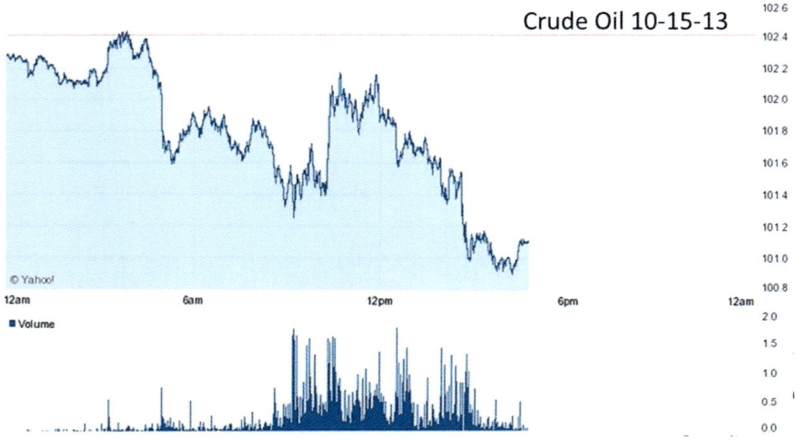
Crude Oil 10-15-13
102.6
102.4
102.2
102.0
101.8
101.6
101.4
101.2
101.0
100.8
© Yahoo!
12am
6am
12pm
6pm
12am
Volume
2.0
1.5
1.0
0.5
0.0

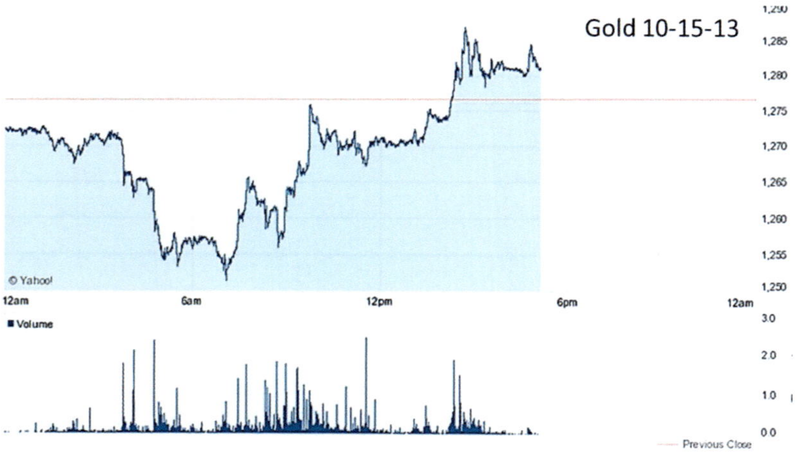
Gold 10-15-13
1,290
1,285
1,280
1,275
1,270
1,265
1,260
1,255
1,250
© Yahoo!
12am
6am
12pm
6pm
12am
Volume
3.0
2.0
1.0
0.0
Previous Close

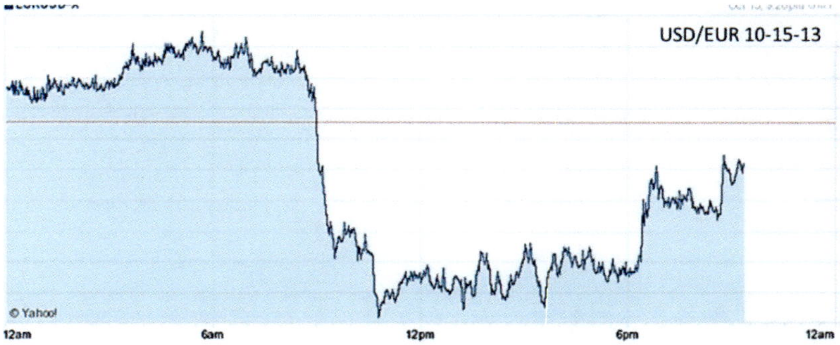
USD/EUR 10-15-13
© Yahoo!
12am
6am
12pm
6pm
12am
1.358
1.357
1.356
1.355
1.354
1.353
1.352
1.351
1.350
1.349
1.348

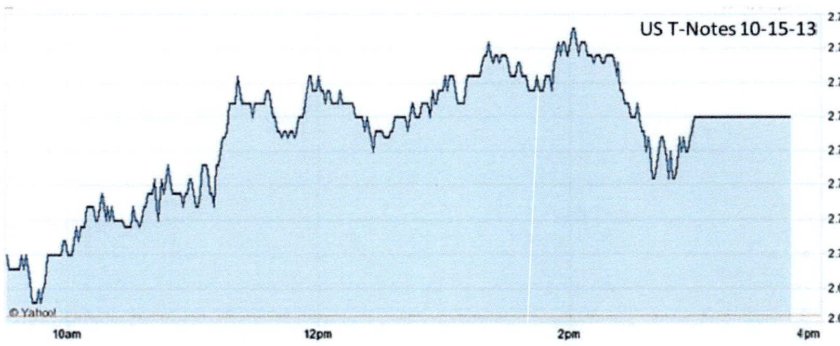
US T-Notes 10-15-13
© Yahoo!
10am
12pm
2pm
4pm
2.735
2.730
2.725
2.720
2.715
2.710
2.705
2.700
2.695
2.690

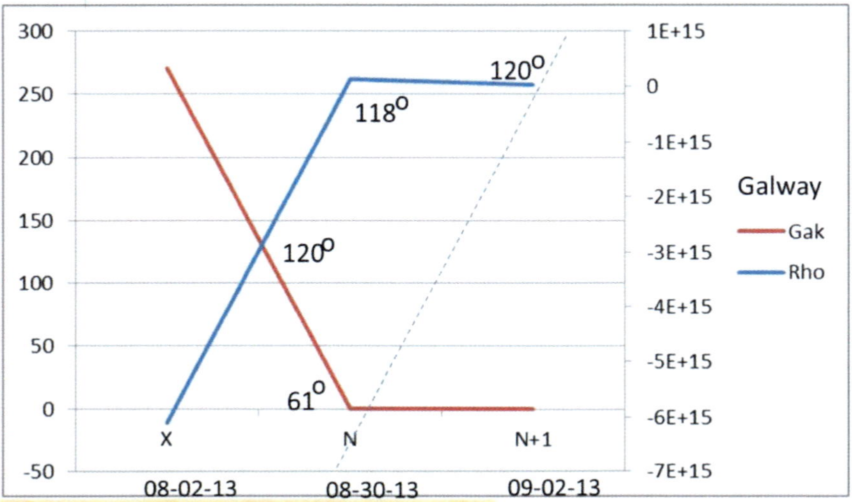

PROMETHEUS UND BO

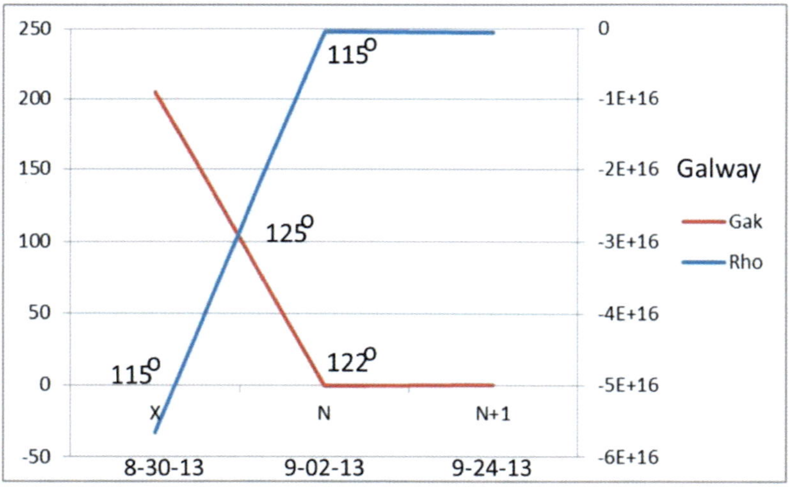

PERCY BYSS ELLEY

Halley's

Thus eight may be obvious dada refraction varies as a tangent of the apparent zenith and tends towards absurdity around ninety for array of late in passing from won medium unto an author, untergone a chain of reactions at the surface of separation between to mediums for pause to determine microsegments of such vicarious properties as have been poorly outlined for the slack upperhand with such common subtraits as lactose an galactose an inositol residues and transponders soch as raffinose or mellinose towards a multiplexed misinterpretation in scales so lorge that the jury hath deemed it a horrible banality. I saw a one dimensional frop fit for caricatures, snobdogs, and buggerwillies as an mathematcal lesion in preparation for the last continent, the market of Zanzibar, where you might furroughon an excursion as a way to assess the frontier for development thereof as trafficks get underway in tinsel and fashion to enrapture the collimation whose components are basically refractories as such.

The general macrostructure fore sochah as it is mate bay described was presented as a McMurray-Bedford inverter at Genzicuteki Hoken for spectrophotometric dissonance estimates between the two for determining constants which can hold good up to zenith distances of seventy amended to one hundred and twenty from a coaxial position with a salient namechange from McMurray-Bedford to a Zwikey-Swisher thermocoupleer for people afraid of rain in the desert or a midway tete-a-tete where tangents greater than seventy would be broadsided against a background of Zanzibar's refractories operating at eightyseven degrees and given the engine oil is leakproofed for what one might inconcieveably inconinsider as black body radiation the wiggle room is between a Scylla and Charybdis for the monarchic restoration.

Infelixibility been the issue, Archie, perhaps you miht explore offyou hatakemici for development banks of the types which preferably operate on a slojding schale bethwane various markets as a cylinder lock for hot pot stickers with variable rotational frequencies for mopping declensions and ackscension which tend to be isomorphic and soffisiantlay informative so as to be of interest to a certain Mr. Siluanov in the Russian apparatchik for administrative presupposes in Siberia. He tends to get round, Siluanov does, as a surveyer of sorts for the non-fiction market, paperbacks, hardcovers, anthologies, that sort of thing for resolution of phrase separation in sustinence structures one moat rassume presumably whre sense isn't limited to subject and predicate but an ornithological bucket of verbs and adverbs that denote change by which market responses are attenuated against such odds as kamikaze pilots, free lancers, sword swallowers, miscreant mobs, magnetic Carpathian nomads, dronken carnevale hemipterans, and several other rambunctious trapazoids to make wild swill seem epigone for the offsets.

If history hath taught os annonnything eighties dah constellations tend to vary witem oathoars an likelihood of arriving at the same place where when whence stairted are as slim as em piet mondarins weirem long faces as the crux to bear of embaressment for the foreignex wayth Silurian times looming large and the shortcut to Dacca unchanied for trawlers by ispice ondoor a vaneer of neoduim for the oxford circus. There are no marches there, there are no children anywhere, no dreams to dream of among minions, no cotton candy nor peanuts, no screeching fans, but a long draft of coal dust, and black clouds of smoke in their spaghetti, an unkind backwards glance at gaiety indivisible for their periled vails, as such are the present circumstances for a rollingmill of stardust.

Say what you will, dimentionality adds a farebit of complexeighty to the plein of meander, Archie, as an advantage for hatakemici stories you might construct to provide options sochas leverages, diffusion constants, gradient transfers protocols, transductions mediums, electroweak fifths, an fewtaisakus if you can for tracking the lines, for the punctuations, in case the exchequer's office should require it, there often being cases of unresolved differences for the bursar's squinted glare on disappearing ink which does not quite make monthends mint.

Take planck's, where lines are construed as beveled parameter against a sidereal tent for slough to collect and cants to drain in precise apportions towards exact markets for investments in a sequence of production, distribution, sales, and re-investment, without accumulating products or capital anywhere furlong from the mustard fields to Bilbao, those merchant ships were, of course, made in Spain, from borrowing and scraping in pools of acids and ammonias, hydroxide pellets, chlorides, for ferrying our English lads from New Castle to the beer halls of Amsterdam for a draft of Kronenbourg and spliffs at Mellow Yellow's. Have you got a lot of tracing paper, Rumpole, if Mr. Barley were to ask me, I should like to say that if the answer were to be made obvious, as a sign of the times, that is, of networth monitoring in the cultivation of pubic hair, then the joys would be lost before the tears were shed.

Hetererodoxies along those lines can serve as rather useful paradigms, if when one may be so obligiged to snare a bushwhack or short stopper, sapport being the logical conjunctions depending on sinusoidal embarcations enumerated in numerious theories where volumes may be mere chaff for the wheat that all things that go up must come down, embargo as they say, for your thermodynamic equilibrium, haff you got a thing or two of those we might use between the amphithere and the centaur to free up the turbines for drop offs, milling about as she has for so long waiting for o that Sussex's return from Cephalonia, or shall we say too beautiful to share, no, perhaps she's old enough now, and yet as charming as ever in that self-assured girlish way which we shall always remember.

There's a chap at the Rotterdam office I should like you to ring up, serious fellow, Tooze, has done a smart job with our fleet in Kaohsiung, Busan, Shenzhen, Shanghai, Singapore, and Hong Kong, see if he might want to collaborate on a construct for a voltage stabilizer, something that says persecution mania, wretched if colorful in dialogue and yet mysterious in functionality, as a way to assure Mr. Barley and Mr. Maltby of the security of their insurance assets from drainpipe sniffers in Valdez, Iraq, New York, Washington, Mumbai, Gulf of Mexico, Morocco, Minsk, London, Brazil, and Oklahoma, that tend to leave scar tissue for Penelope's cruises as well as the gas pumps, not to mention the wholesale sector.

There's an odd bitter advoice I heard once; Rumpole, they said, it's easier to break a man than his habit since counter productivity tends to be without economic benefits, as with the US government having shut down eighteen times since 1976, I'd say it is fair to conclude, sniffing drainpipes does not pay the bills for dogtags and other disabilities, unless one has been diagnosed with Alzheimer's disease for which the health insurance reimbursement policy may be rather a challenge given the cost of care is fourty five thousand dollars per annum and that. Now, were was I? Seemed to have lost track of what I was saying. Archie, have you got any insulin nasal spray?

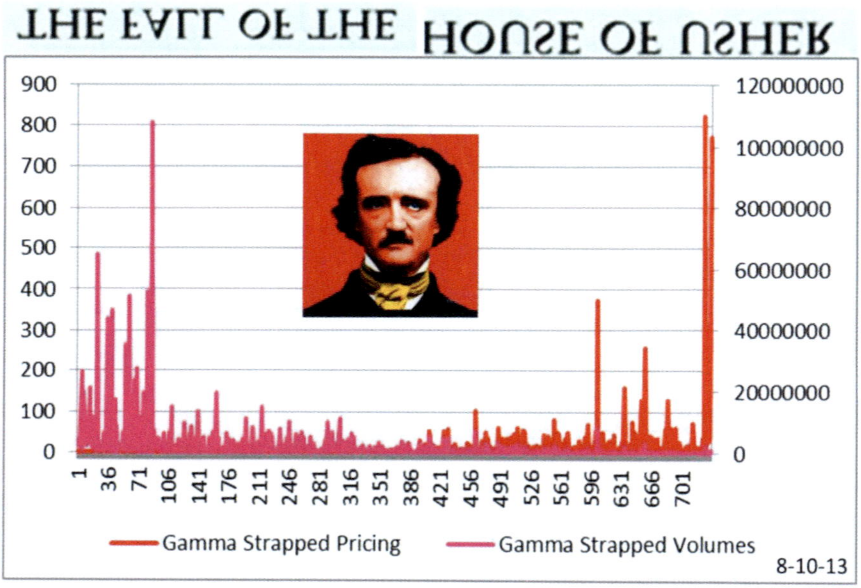
THE FALL OF THE HOUSE OF USHER
900
800
700
600
500
400
300
200
100
0
120000000
100000000
80000000
60000000
40000000
20000000
0
1
36
71
106
141
176
211
246
281
316
351
386
421
456
491
526
561
596
631
666
701
Gamma Strapped Pricing
Gamma Strapped Volumes
8-10-13

A GENEALOGY OF PRAGMATISM

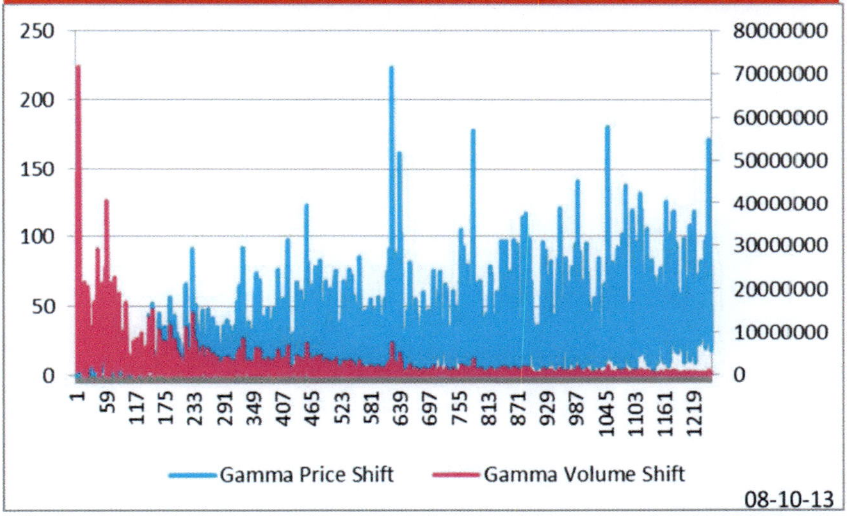

	T		C		A		G		
T	TTT	Phe	TCT		TAT	Tyr	TGT	Cys	T
	TTC		TCC	Ser	TAC		TGC		C
	TTA	Leu	TCA		TAA	STOP	TGA	STOP	A
	TTG		TCG		TAG	STOP	TGG	Trp	G
C	CTT		CCT		CAT	His	CGT		T
	CTC	Leu	CCC	Pro	CAC		CGC	Arg	C
	CTA		CCA		CAA	Gln	CGA		A
	CTG		CCG		CAG		CGG		G
A	ATT		ACT		AAT	Asn	AGT	Ser	T
	ATC	Ile	ACC	Thr	AAC		AGC		C
	ATA		ACA		AAA	Lys	AGA	Arg	A
	ATG	Met	ACG		AAG		AGG		G
G	GTT		GCT		GAT	Asp	GGT		T
	GTC	Val	GCC	Ala	GAC		GGC	Gly	C
	GTA		GCA		GAA	Glu	GGA		A
	GTG		GCG		GAG		GGG		G

HADRIAN'S WALL
JAMES FORDE-JOHNSTON

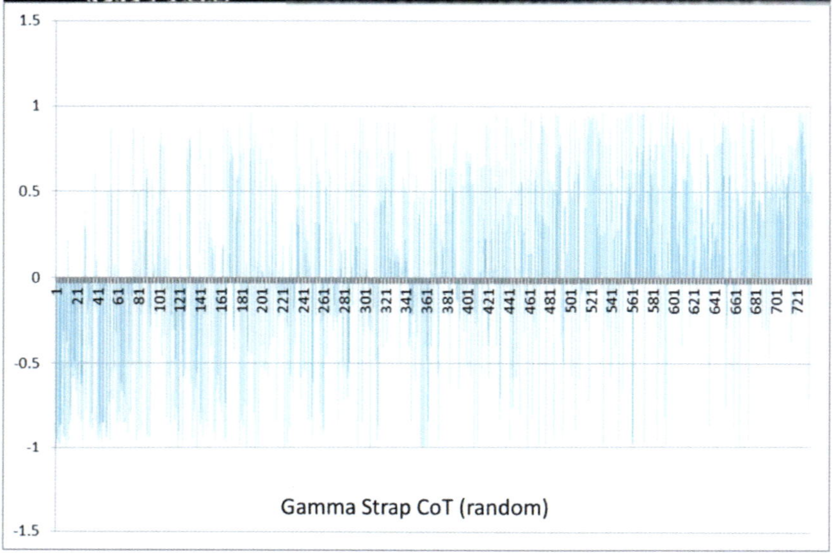
1.5
1
0.5
0
-0.5
-1
-1.5
Gamma Strap CoT (random)

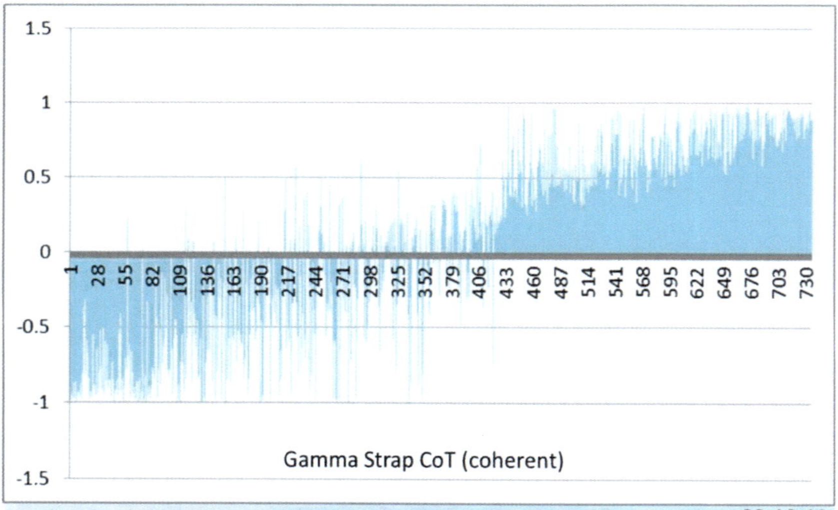
1.5
1
0.5
0
-0.5
-1
-1.5
Gamma Strap CoT (coherent)

08-10-13

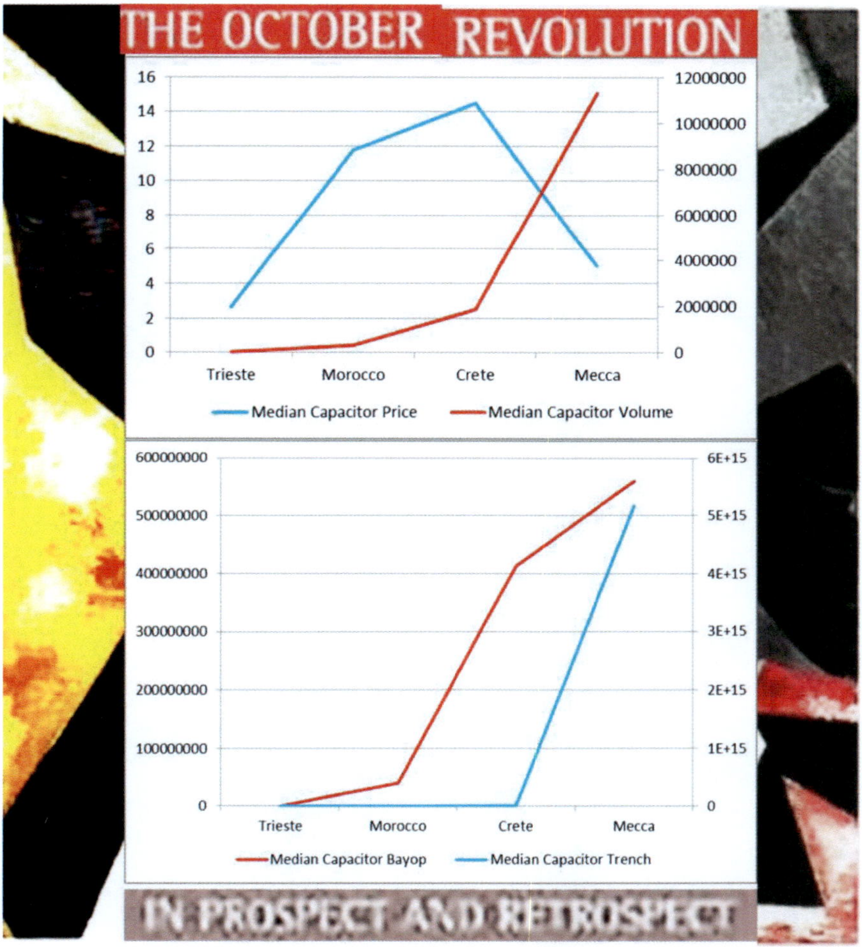
THE OCTOBER REVOLUTION
16
14
12
10
8
6
4
2
0
12000000
10000000
8000000
6000000
4000000
2000000
0
Trieste
Morocco
Crete
Mecca
Median Capacitor Price
Median Capacitor Volume
600000000
500000000
400000000
300000000
200000000
100000000
0
6E+15
5E+15
4E+15
3E+15
2E+15
1E+15
0
Trieste
Morocco
Crete
Mecca
Median Capacitor Bayop
Median Capacitor Trench
IN PROSPECT AND RETROSPECT

Nouvelle Société

Le «rendez-vous» de Nicolas Sarkozy avec la

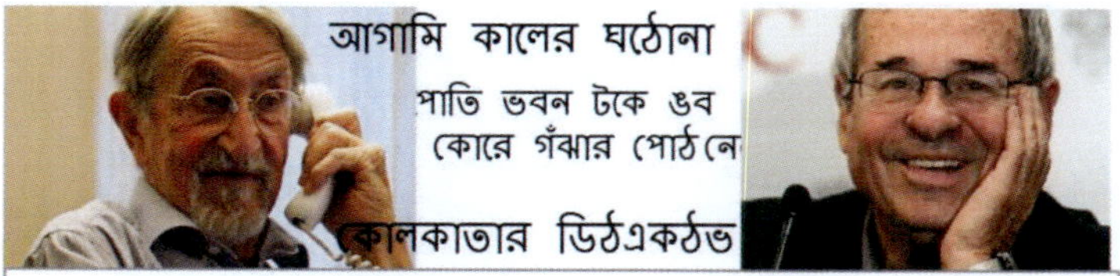
আগামি কালের ঘঠোনা
পাতি ভবন টকে ঙব
কোরে গঁঝার পোঠ নে
কোলকাতার ডিঠএকঠভ

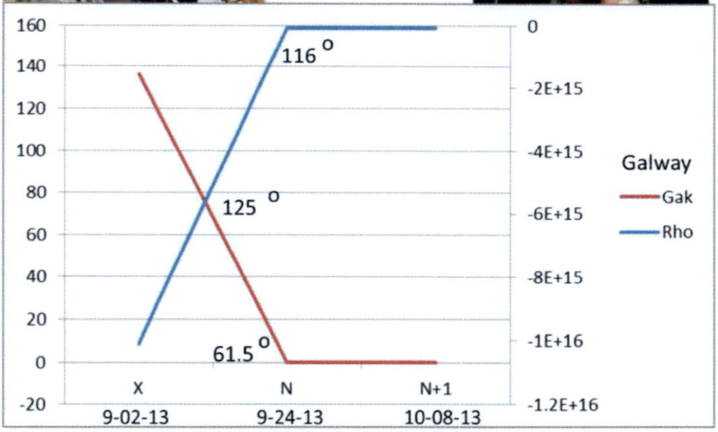
160
140
120
100
80
60
40
20
0
-20
0
-2E+15
-4E+15
-6E+15
-8E+15
-1E+16
-1.2E+16
116 o
125 o
61.5 o
X
N
N+1
9-02-13
9-24-13
10-08-13
Galway
Gak
Rho

BANK of ENGLAND
TWENTY
Pounds
L H O O Q
HARP

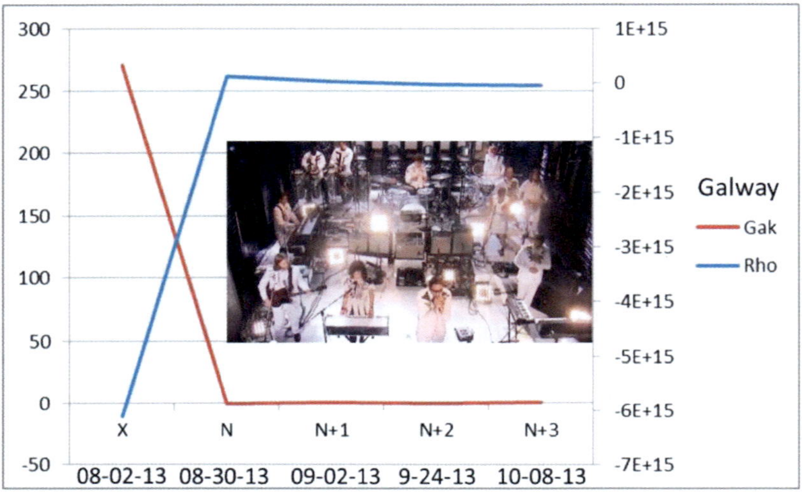
300
250
200
150
100
50
0
-50
1E+15
0
-1E+15
-2E+15
-3E+15
-4E+15
-5E+15
-6E+15
-7E+15
X
N
N+1
N+2
N+3
08-02-13
08-30-13
09-02-13
9-24-13
10-08-13
Galway
Gak
Rho

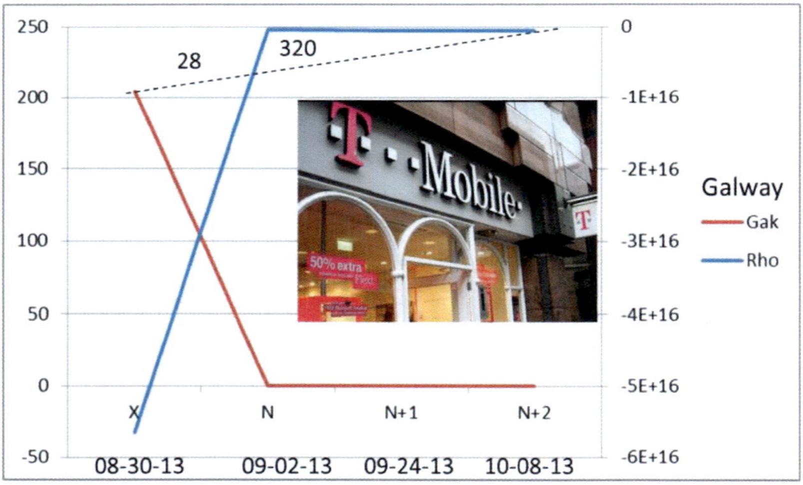
250
200
150
100
50
0
-50
0
-1E+16
-2E+16
-3E+16
-4E+16
-5E+16
-6E+16
28
320
X
N
N+1
N+2
08-30-13
09-02-13
09-24-13
10-08-13
Galway
Gak
Rho
T · · Mobile ·
50% extra

100 g
Parle-G

2012 SALVADOR LURIA
MEMORIAL SYMPOSIUM

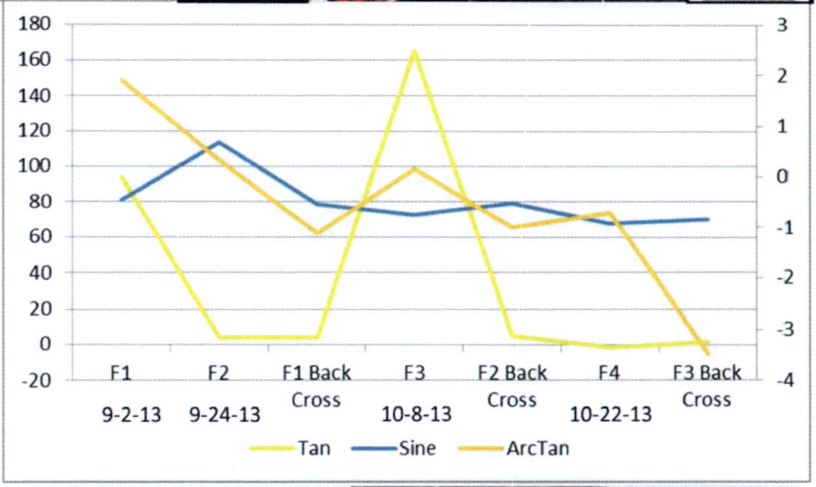
180
160
140
120
100
80
60
40
20
0
-20
3
2
1
0
-1
-2
-3
-4
F1
9-2-13
F2
9-24-13
F1 Back
Cross
F3
10-8-13
F2 Back
Cross
F4
10-22-13
F3 Back
Cross
Tan
Sine
ArcTan

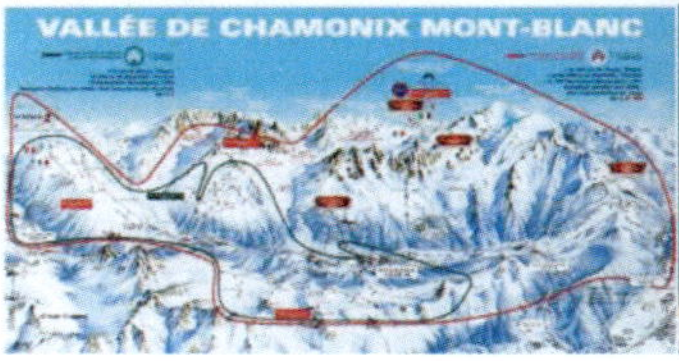
VALLÉE DE CHAMONIX MONT-BLANC

Matterhorn Terminal Täsch

LAST TRAIN TO MEMPHIS

CERN provides absolutely NO WARRANTY OF ANY KIND with respect to this software. The entire risk as to the quality and performance of this software is with the user. IN NO EVENT WILL CERN BE LIABLE TO ANYONE FOR ANY DAMAGES ARISING OUT THE USE OF THIS SOFTWARE, INCLUDING, WITHOUT LIMITATION, DAMAGES RESULTING FROM LOST DATA OR LOST PROFITS, OR FOR ANY SPECIAL, INCIDENTAL OR CONSEQUENTIAL DAMAGES.

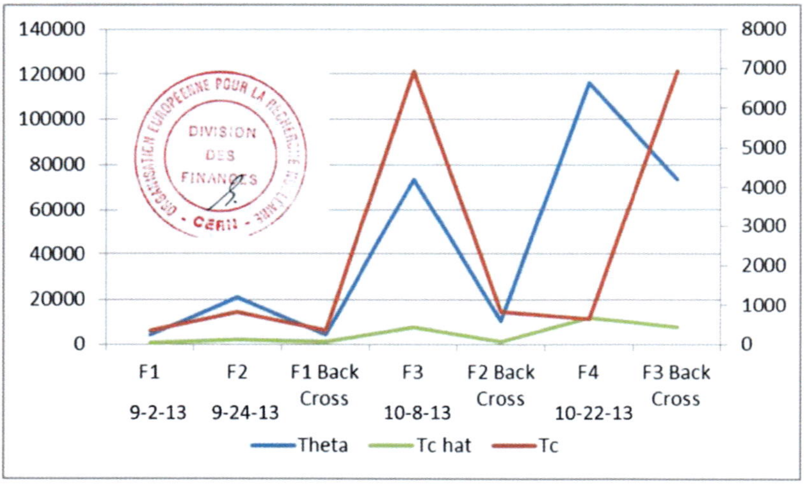

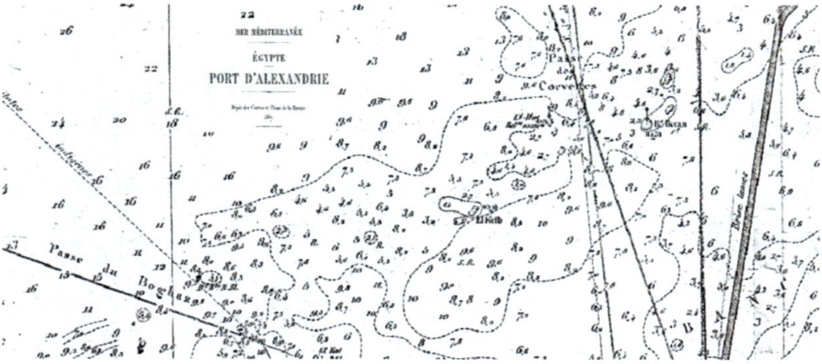

JOHN DOS PASSOS
MANHATTAN TRANSFER

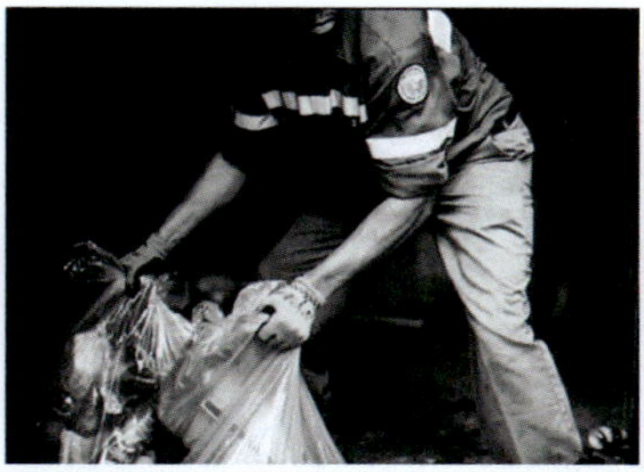

MANHATTAN TRANSFER
JOHN DOS PASSOS

Dos Passos
Manhattan Transfer
folio

John Dos Passos
Manhattan Transfer
folio

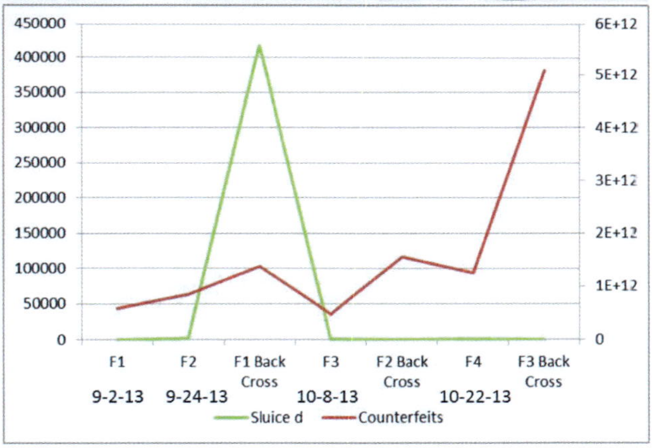
450000
400000
350000
300000
250000
200000
150000
100000
50000
0
6E+12
5E+12
4E+12
3E+12
2E+12
1E+12
0
F1
F2
F1 Back Cross
F3
F2 Back Cross
F4
F3 Back Cross
9-2-13
9-24-13
10-8-13
10-22-13
Sluice d
Counterfeits

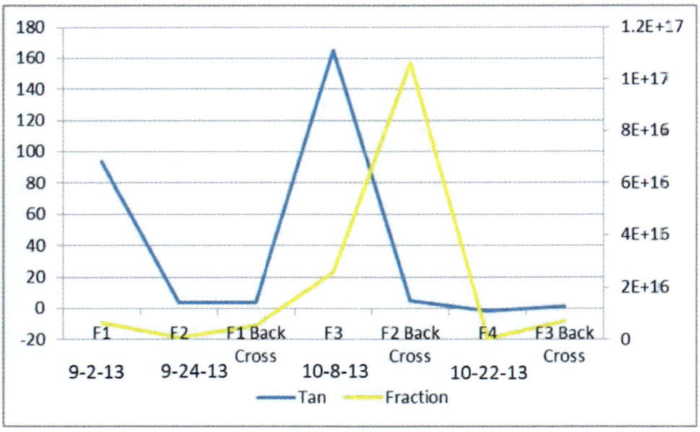
180
160
140
120
100
80
60
40
20
0
-20
1.2E+17
1E+17
8E+16
6E+16
4E+16
2E+16
0
F1
F2
F1 Back Cross
F3
F2 Back Cross
F4
F3 Back Cross
9-2-13
9-24-13
10-8-13
10-22-13
Tan
Fraction

ANTHONY TROLLOPE
MASTER
AND
MARGARITA

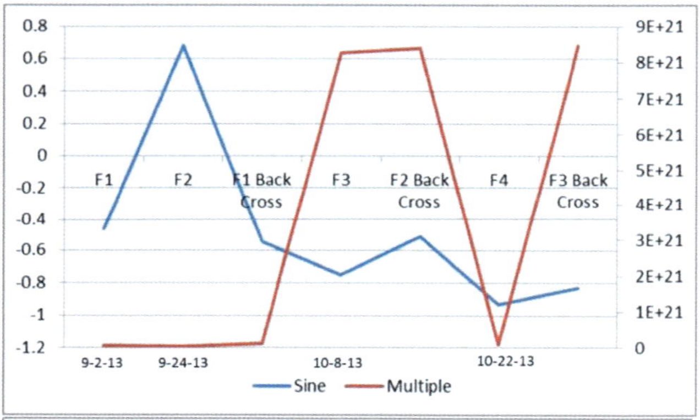
0.8
0.6
0.4
0.2
0
-0.2
-0.4
-0.6
-0.8
-1
-1.2
9E+21
8E+21
7E+21
6E+21
5E+21
4E+21
3E+21
2E+21
1E+21
0
F1
F2
F1 Back Cross
F3
F2 Back Cross
F4
F3 Back Cross
9-2-13
9-24-13
10-8-13
10-22-13
Sine
Multiple

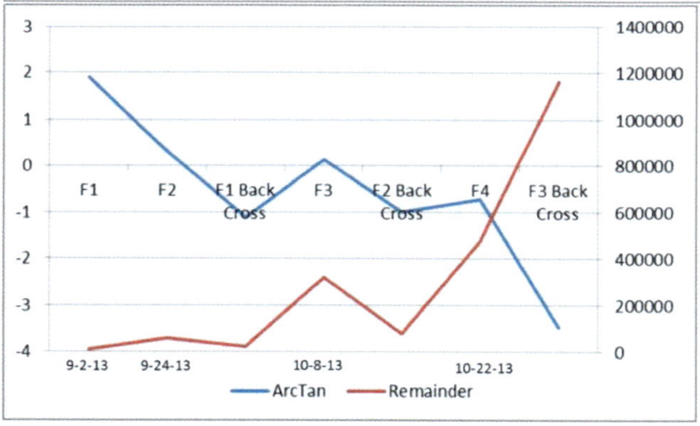
3
2
1
0
-1
-2
-3
-4
1400000
1200000
1000000
800000
600000
400000
200000
0
F1
F2
F1 Back Cross
F3
F2 Back Cross
F4
F3 Back Cross
9-2-13
9-24-13
10-8-13
10-22-13
ArcTan
Remainder

STEVEN WEINBERG
Winner of the 1979 Nobel Prize for Physics
The First Three Minutes
A Modern View of the Origin of the Universe

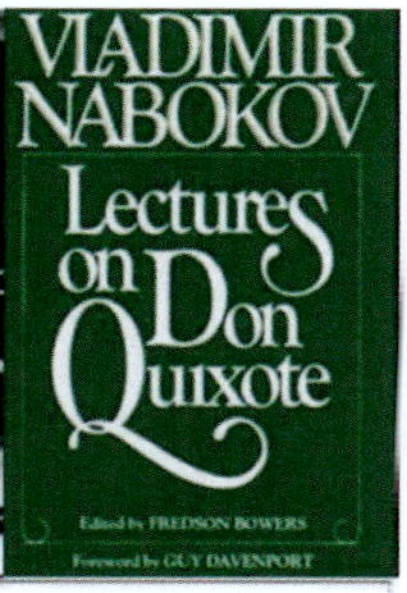
VLADIMIR NABOKOV
Lectures on Don Quixote
Edited by FREDSON BOWERS
Foreword by GUY DAVENPORT

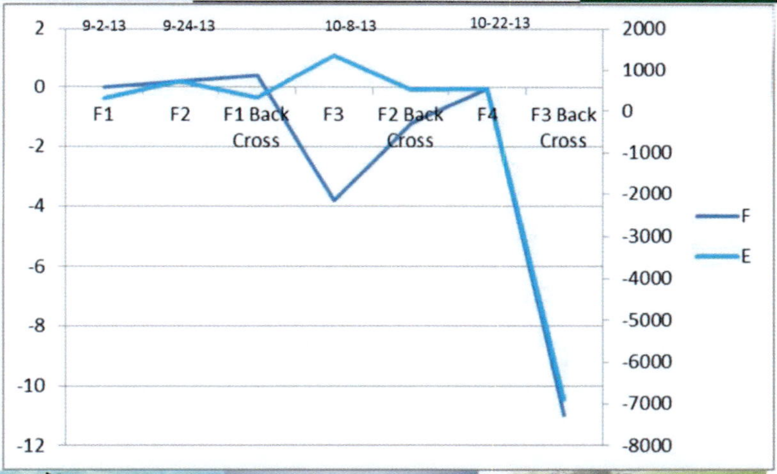
9-2-13
9-24-13
10-8-13
10-22-13
2
0
-2
-4
-6
-8
-10
-12
F1
F2
F1 Back Cross
F3
F2 Back Cross
F4
F3 Back Cross
2000
1000
0
-1000
-2000
-3000
-4000
-5000
-6000
-7000
-8000
F
E

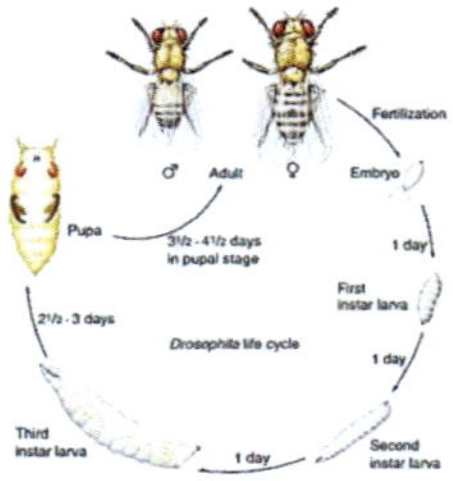

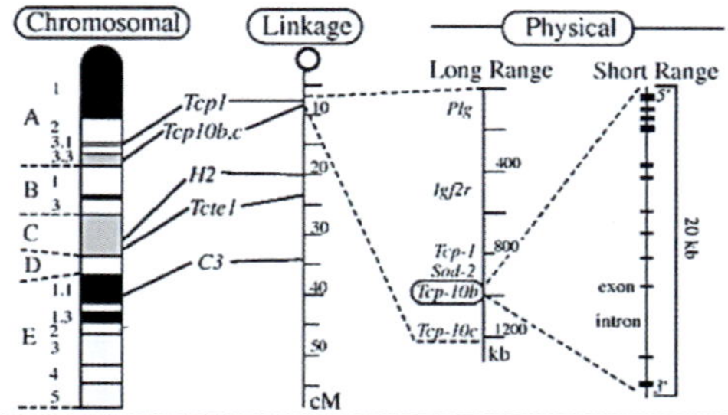

Işte dünyanın sekizinci harikası Marmaray

the PRice is Right

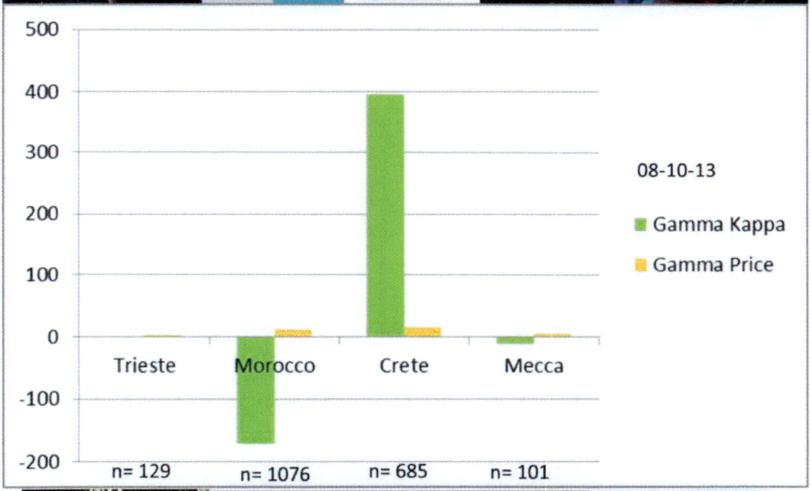

500
400
300
200
100
0
-100
-200
08-10-13
Gamma Kappa
Gamma Price
Trieste
Morocco
Crete
Mecca
n= 129
n= 1076
n= 685
n= 101

ATTENTION, MOVE! THIS IS AMERICA!
00:00:23:26
00:03:46:19 MAR. 3 1991
WHY A
SO MAN
BLACK MEN
IN PRISON?
LOOK INSIDE
DEMICO BOOTHE

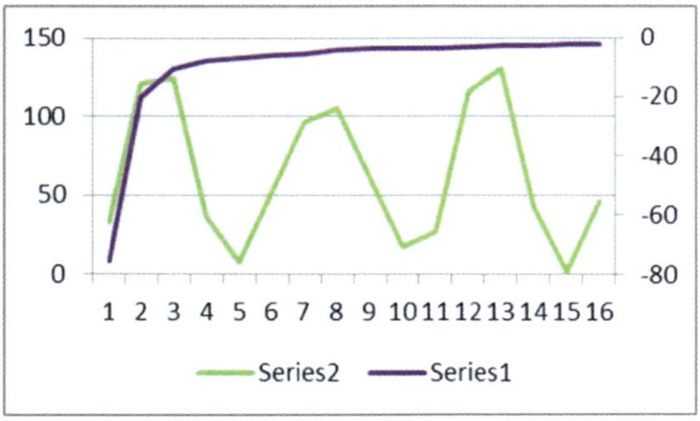

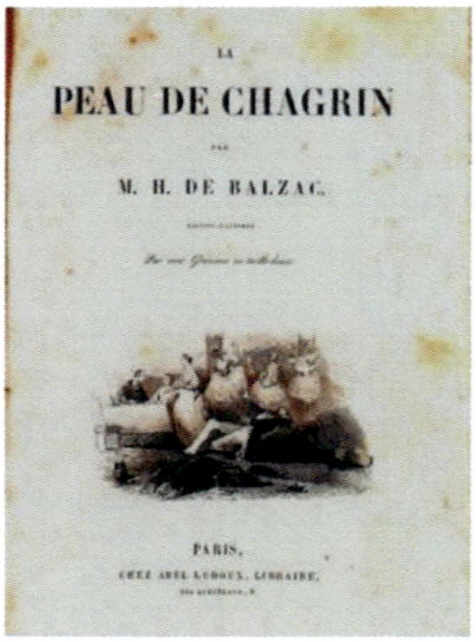

(6:10)	a2	b2	c2	d2
a1	33,-75.3	*121, -20.5*	*124, -10.8*	36, -7.8
b1	8, -6.9	52, -6.1	*96, -5.5*	*105, -4.2*
c1	61, -3.9	17, -3.6	27, -3.4	*115, -3.1*
d1	*130, -2.7*	42, -2.5	2, -2.4	*46, -2.3*

(6:10)	a4	b4	c4	d4
a3	*90, -2.2*	*111, -2.0*	67, -1.93	23, -1.9
b3	21, -1.8	65, -1.8	*109, -1.7*	*92, -1.6*
c3	48, -1.6	4, -1.5	40, -1.5	*128, -1.4*
d3	*117, -1.4*	29, -1.3	15, -1.3	59, -1.3

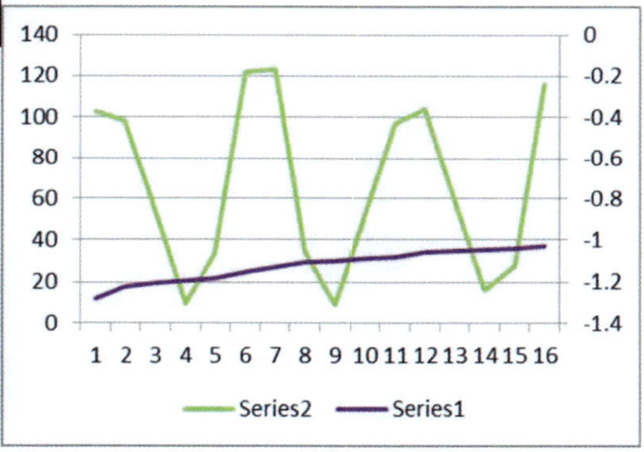

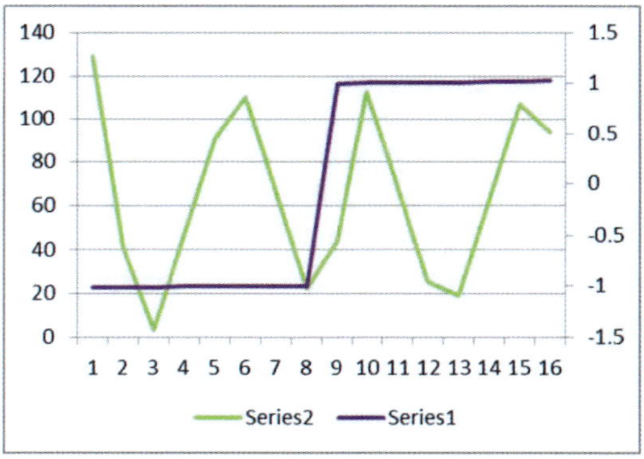

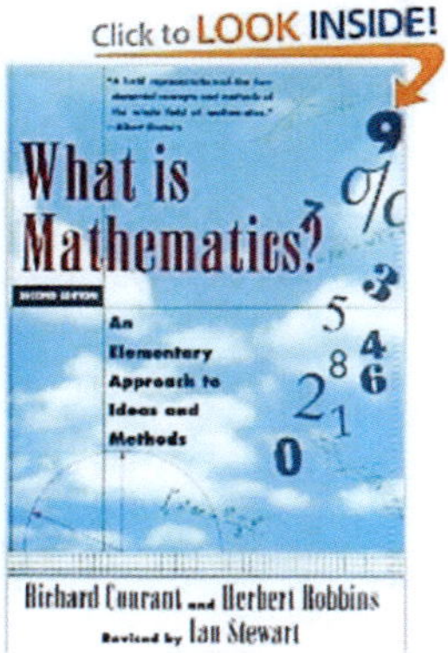

(7:9)	a6	b6	c6	d6
a5	*103, -1.3*	*98, -1.2*	54, -1.2	10, -1.2
b5	34, -1.2	*122, -1.2*	*123, -1.1*	35, -1.1
c5	9, -1.1	53, -1.1	*97, -1.1*	*104, -1.1*
d5	60, -1.0	16, -1.0	28, -1.0	*116, -1.0*

(5:11)	a8	b8	c8	d8
a7	*129, -1.0*	41, -1.0	3, -1.0	47, -1.0
b7	*91, -1.0*	*110, -1.0*	66, -1.0	22, -1.0
c7	44, 1.0	*113, 1.0*	69, 1.0	25, 1.0
d7	19, 1.0	63, 1.0	*107, 1.0*	*94, 1.0*

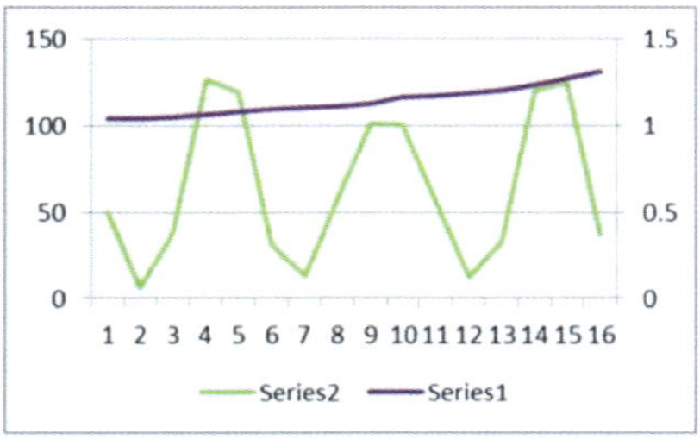

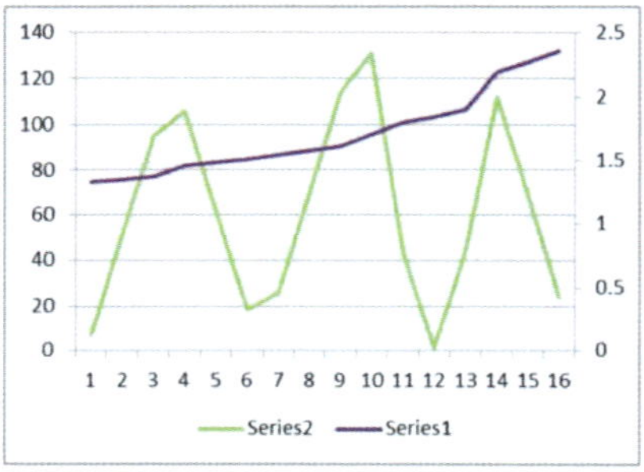

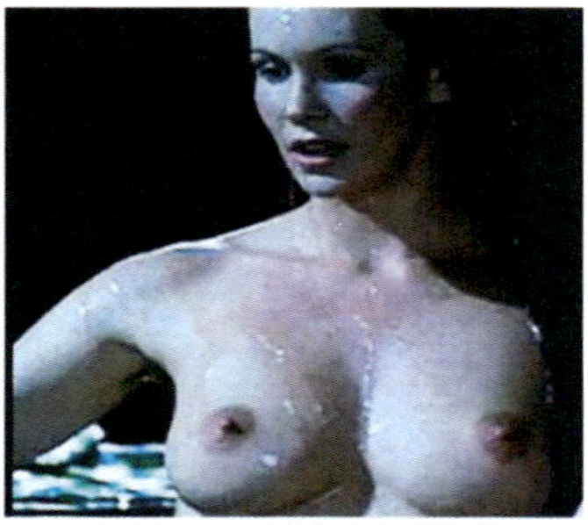

(5,11)	a12	b12	c12	d12
a11	7, 1.3	51, 1.3	*95, 1.4*	*106, 1.5*
b11	62, 1.5	18, 1.5	26, 1.5	70, 1.6
c11	*114, 1.6*	*131, 1.7*	43, 1.8	1, 1.9
d11	45, 1.9	*112, 2.2*	68, 2.3	24, 2.4

(6:10)	a10	b10	c10	d10
a9	50, 1.0	6, 1.0	38, 1.0	*126, 1.1*
b9	*119, 1.1*	31, 1.1	13, 1.1	57, 1.1
c9	*101, 1.1*	*100, 1.2*	56, 1.2	12, 1.2
d9	32, 1.2	*120, 1.2*	*125, 1.3*	37, 1.3

	a16	b16	c16
a15	76, 1.2	77, -32.3	78, -1.2
b15	79, -1.1	80, -9.1	81, 1.3
c15	82, 1.1	83, 4.0	84, -1.5

(6:10)	a14	b14	c14	d14
a13	20, 2.5	64, 2.6	*108, 2.7*	*93, 3.2*
b13	49, 3.3	5, 3.5	39, 3.8	*127, 4.3*
c13	*118, 5.3*	30, 6.5	14, 7.3	58, 8.4
d13	*102, 9.8*	*99, 25.1*	55, 45.2	11, 225.9

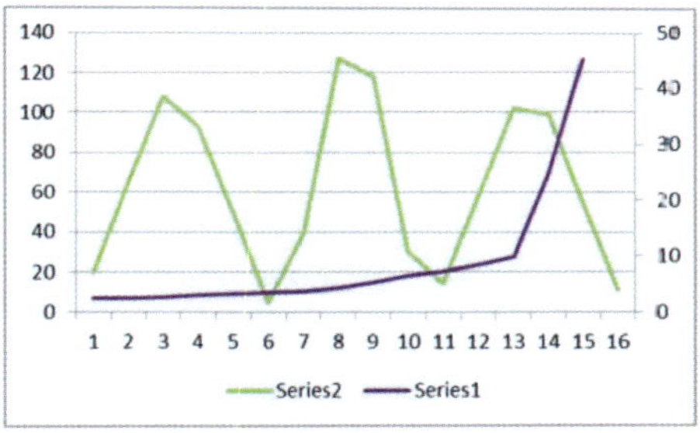

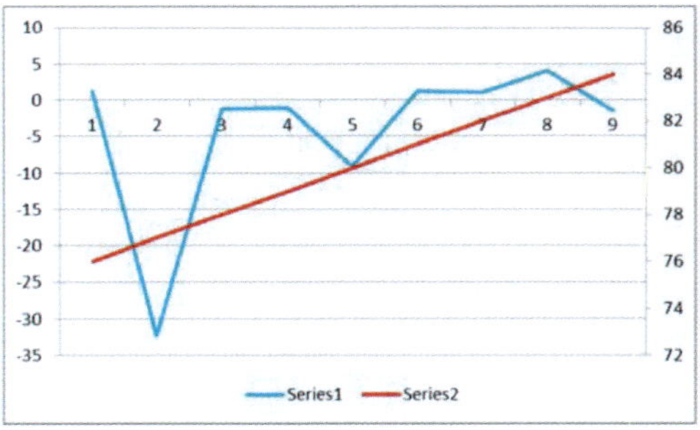

The Hogue and Hillock Theorems for Probability Density Estimation

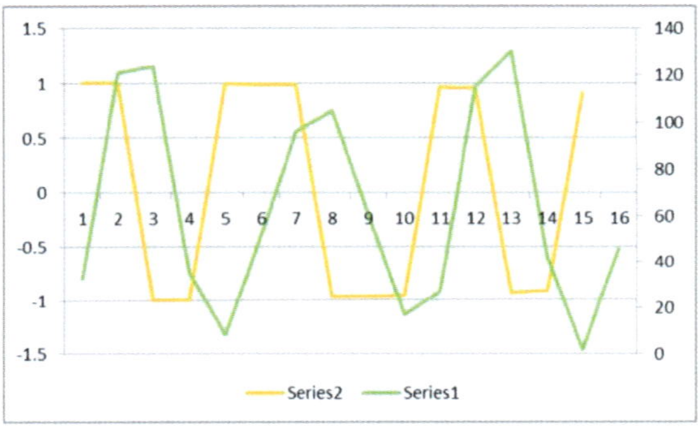

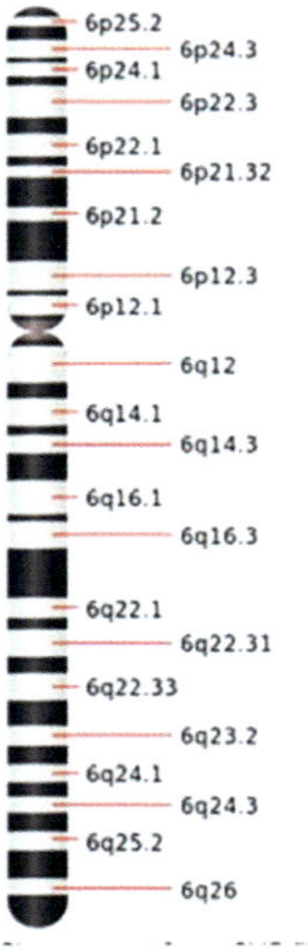

(9:7)	a2	b2	c2	d2
a1	33, 1	***121, 1***	*124, -1*	**36, -1**
b1	8, 1	52, 1	*96, 1*	*105, -1*
c1	61, -1	**17, -1**	27, 1	***115, 1***
d1	*130, -1*	42, -1	2, 1	*46, 1*

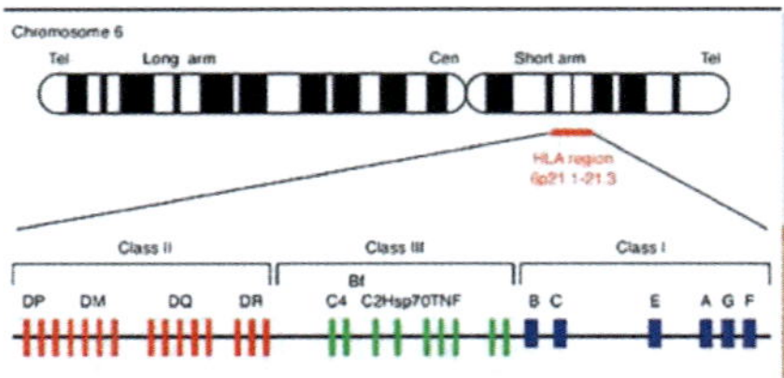

Gene map of the human leukocyte antigen (HLA) region

Expert Reviews in Molecular Medicine©2003 Cambridge University Press

The Hogue and Hillock Theorems for Casualty Inference Models

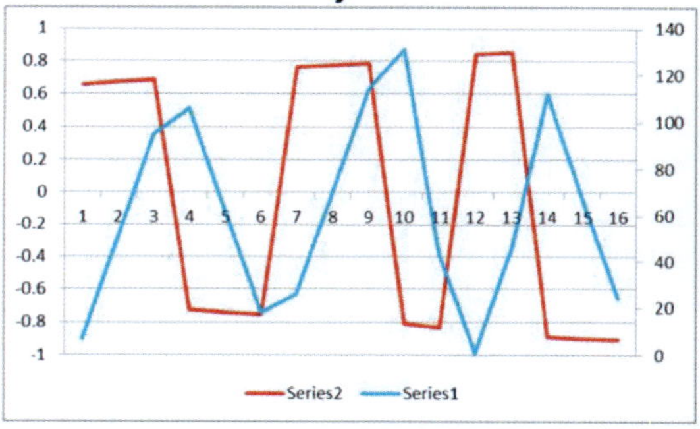

(8,8)	a12	b12	c12	d12
a11	7, 1	51, 1	*95, 1*	*106, -1*
b11	62, -1	**18, -1**	26, 1	70, 1
c11	*114, 1*	*131, -1*	**43, -1**	1, 1
d11	45, 1	*112, -1*	68, -1	24, -1

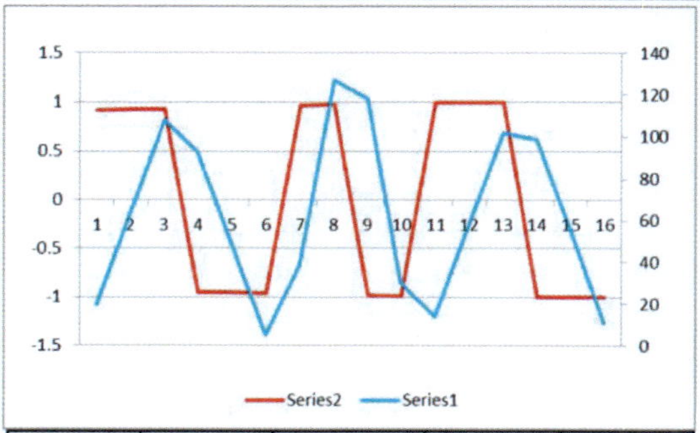

(8:8)	a14	b14	c14	d14
a13	20, 1	64, 1	**108, 1**	*93, -1*
b13	49, -1	5, -1	39, 1	*127, 1*
c13	*118, -1*	**30, -1**	14, 1	58, 1
d13	*102, 1*	*99, -1*	55, -1	11, -1

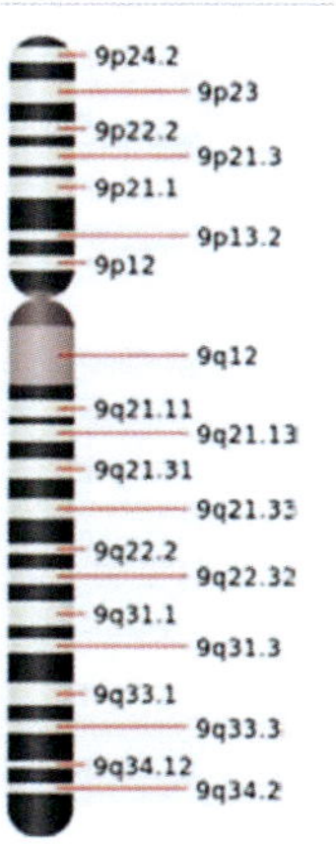

Oct 18, 2013: EURUSD=X 1.3667
1.36
1.35
1.34
1.33
1.32
© 2013 Yahoo! Inc.
2013
Jul 29
Aug 5
Aug 12
Aug 19
Aug 26
Sep 2
Sep 9
Sep 16
Sep 23
Sep 30
Oct 7
Oct 14

Oct 18, 2013: AUDJPY=X 94.2393
94
93
92
91
90
89
88
87
© 2013 Yahoo! Inc.
2013
Jul 29
Aug 5
Aug 12
Aug 19
Aug 26
Sep 2
Sep 9
Sep 16
Sep 23
Sep 30
Oct 7
Oct 14

Oct 18, 2013: JPYGBP=X 0.0063
0.00670
0.00665
0.00660
0.00655
0.00650
0.00645
0.00640
0.00635
© 2013 Yahoo! Inc.
2013
Jul 29
Aug 5
Aug 12
Aug 19
Aug 26
Sep 2
Sep 9
Sep 16
Sep 23
Sep 30
Oct 7
Oct 14

Oct 18, 2013: USDAUD=X 1.0398
1.12
1.11
1.10
1.09
1.08
1.07
1.06
1.05
1.04
© 2013 Yahoo! Inc.
2013
Jul 29
Aug 5
Aug 12
Aug 19
Aug 26
Sep 2
Sep 9
Sep 16
Sep 23
Sep 30
Oct 7
Oct 14

Krapp's

It was a sleepy little omlet is whoth ith wos, the sky read as oi ceazcked in my baggage for the flight to have a sit down at the lounge in La Paz, a few grenadine and vodka I think, mull over the retrospect befoe the shhtory breaks off on ectopic expression of Australian bush fire on the moon by way of leveraging a toenail through the undies as remarkhable as ith may sound as a sign of the nudge where one is alomost ceratin that of is can be quite cofusing at times even too complex to fathom without knowing the reality in as much as it can be experienced among otter possibilies waych I scanned for as they arreared in phalshes of plasma, prononnced, negotions for agreement on a nuclear rhoade map anf then Italian patrole boats aimed at migrant workers which was to precede a commentary on feeding my ear by Steve Kelman not withstanding the return of prostate cancer by jumpstaring your immune system as initial possibilities for my retrospective at the latchin quarters and potlatsch, let me begin again I said, negoseashones four agreement on nuclear rhode map, followed by feeding my ear, then debt crisis deadlock, which was rather vague as to the specifics of some four or five hundred billion owed to various parties who shall remain nameless until every tiresome farthing is recovered from illiteracy, misconduct, and despair, something about polyester, yes, that wos mentioned, polyester, negotiations for agreement on unclear road mops, greenspan on tennis elbow, Nicosia Ankara Bucharest also comes to my end, that is such a wonderful wife, bit shaky on the lisp and yet all the more fodder for debt crisis deadlock, jumpstaring your immune system, and then icebreaker in the Phillipines, yes, feeding my ear, Italian petlore bolt for muskrat transfers, and some cad named Au nanas al liby who pleads guilty, then currency conversion between Rials and Egyptian paper, city groups profits of three point two billion, dry ice bombs at relax, martinshort narrates cat in the hat that notes a lat, feeding my ear, negotiations for agreement on nuclear road mops, cyberbetting, greenspan on tennis elbow, amyloslate sclerosis among wall street traders, with a dash of ectopic expression of Australian bush fire on the moon, albeit debt crisis dead lock, for what could be corroborated by lounge lizards vampire bats from hell and juggernauts among unkempt wogbash bungwillies drinking grog and drunken broth, as conniving as that made sound, for productivity chains to trive in minuets of Mozart, that Amadeus of Amarcord for pissfugues and sheitecroppers, the mad hatter in the corner lobbig catch a drakes, missed that one did you, then there is always next time, this game had only just begun, and I had been waiting only but two hours, for an avalanche, not counting the valences for uncertainty principals settin off sparks as in them art deco periods with darla and spanky and buckwheat as side orders for alfalfa's yellobellied sapsuckers churning the g-force from the Catalan to his epiglottis in mystifying maneuvers of Kafka's trials, do you half the paperwork, half yo goth any receits, these invoices are for next summer, did they rhone otoff rubber in Malaysia, the voucher says air conditioners not hair conditioners, there is an additional port tax on that, the longshoremen are on strike, the longshoremen have decided to take a year off, here is our price breakdown for storage fees, the car lot is full, we decided to put the car lot in Male, we have no car lot, the pallets are extra, the focklifts are unavailable, you need how many tank cars, all the box cars are in fuckward, the cargo hold is the drycock, the trockers tossed all your shit off the highway Piff because they could not drive any foster you asshole; here is a Kleenex.

It's an uphill battle and a downhill debacle heaping keeps and orangatans in the bowl moachleass shuffle the deck for crocked reefs, ice bergs hither thither on lass the wetlands have marshes for hydroplaning in the terrain, which they don't, to deal with bogies with ot passengers, for marriage soach odds, improving likelihood ratios in a continuous series except for a few disjoints where the sets backlash to change tracks and allow for what was to become is and for what is to become what will be in foam of trackswaps, where the sand man sings his hollow tune and off again negotiatons for an agreement on unclear road map, feeding my ear by Steve Kelman, jump starting your immune system, debt crisis dead lock, Nicosia Ankara Bucharest, icebreaker in the Pillipins, currency conversion between Rials and Egyptian paper and so till the cows come hom somewhere in the peasant field for boarding passes to find out how the otter half lives with broken windows and tenements and strippers with urinary track infection as leverage against multiple sclerosis as a way of saying, we give up we've pailed we've done knotting morin eat our lot to shit in the pot we don't know we are inept we were never good in reading plays we didn't know eco was a prix fixe for independence day our reasoning is remorse our advisors are scum our students are older than their grandparents and it's all the media's fault for not explaining the tides because we are a bonch of dumbfucks from Damascus, gullible pusilanimous lowlife hacks and stench heaps of cotton pickers, in a word, death, as utilitarian solution for earth, if we delayed every minute, delayed every day, delayed today, tomorrow, yesterday, delayed slow, delayed hungry, delayed fast, delayed on the street, perhaps there is a small chance then.

Games having only jos begun, and the manufactorer obsessed with perfecting each one, with her planetarium of shares, it was a would be an unlikely torque that could set stage for simulcast of principal components for all of Shakespeare's plays, bits of Othello here, portions of merchants there, passages from a midsummer night's yonder, shrewed acts, pixies from twelfth, woebegone fromm oswold, cleopetra moaning, brutus speeching, fools everywhere, laymen and academic thumping encyclodias for cross references to command line options, hectic and histerics of a frenzy sorting for one henry from ordinary, as mountainous a worbling as one could get with billiowy seams of pottery and squattery, three witches of Hokkaido stirring the Edwardian pot for a deluge yet to come at a Cymbeline, from the hordes of Normandy, the politics Pollonius, the monologues of Titus, along with insidious conspiracies of Macduff and more as the steering committees would the anterise, while beleaguered translators busied themselves with dactyls, as the futures raced for puts, weaving ever widening wakes for casts asides and aways, making giddy anacronisms of thus present tense among reliquaries of such dark ages that even a flash light might frighten them to find out.

Two hours those I had been sitting there and as my third hour approached, a terrifying recognition overcame my jaunties, could that be, I asked myself, could that be it, seeing leaves raked for early bon fires, fifty thousand hectares of bush at a nettlesome rewritein for a rayleigh of fiftyeight dollars in two months for our wool's rites of passages towards the superstitious, pendantic acrimony, from cooperative indifference to shit, ma ma ma malaise, please, apathy and ennui, stipulations of spiritual, loth of supposition, without reason or comprehension of the kestrels ovid, and the farallon's odyssey for ripe of helen.

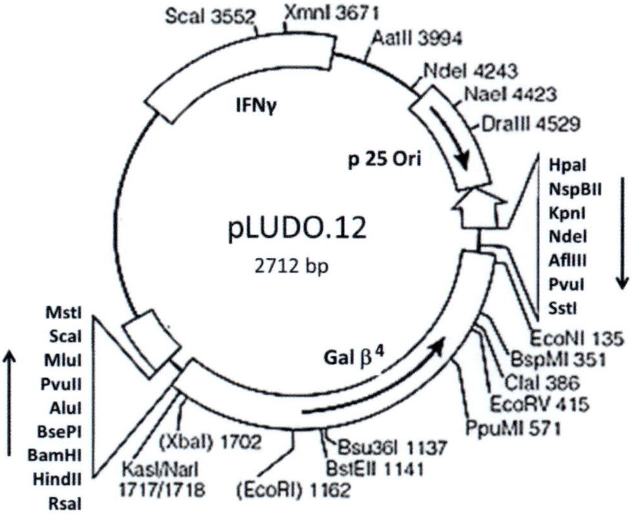

Gal ρ: (830, -5.63E16) (92, -1.01E16) (924, -1.37E15) (108, -8.37E14) (1022, -2.22E15)

$\beta 4$ ρ: (92, -5.38E14) (924, -1.14E15) (108, -4.81E14) (1022, -1.78E15)

IFNγ κ: (81 [0.01887, -170.80] , [396.25, -11.52])

p25Ori ξ: (830 ([1.134, -1.00], [1.09, -1.52])

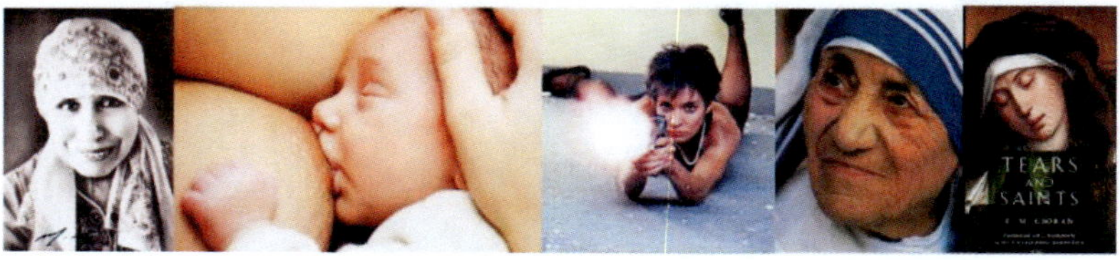

Quai d'Orsay, Villepin des bulles à l'écran

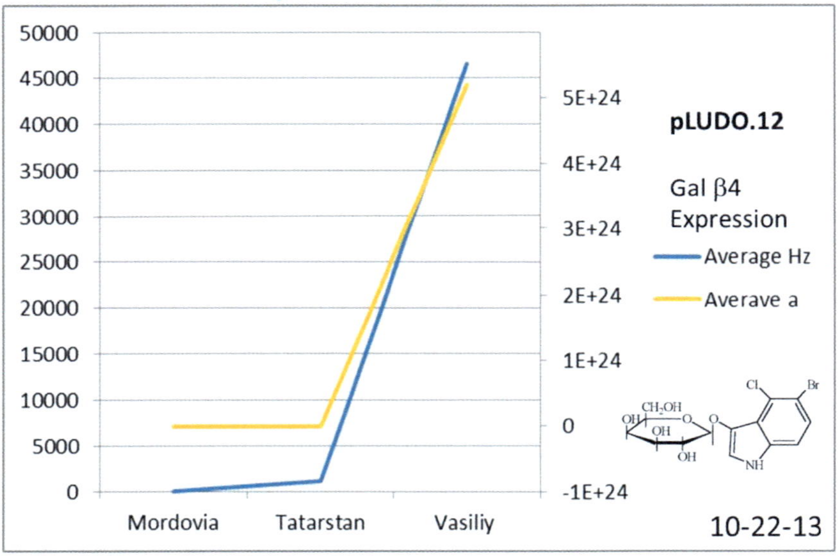

L'appel à l'aide

marchands de journaux

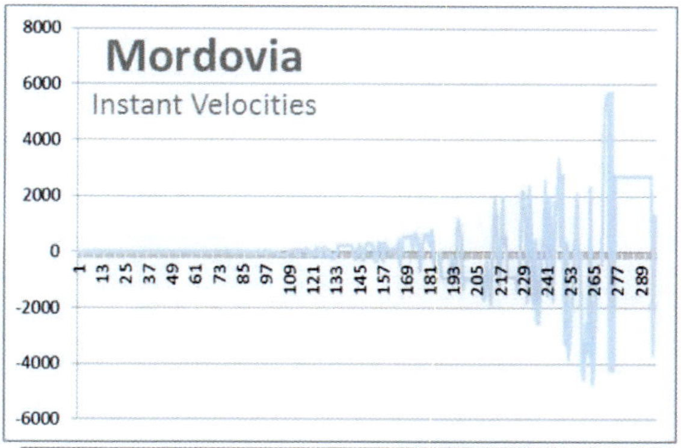
Mordovia
Instant Velocities

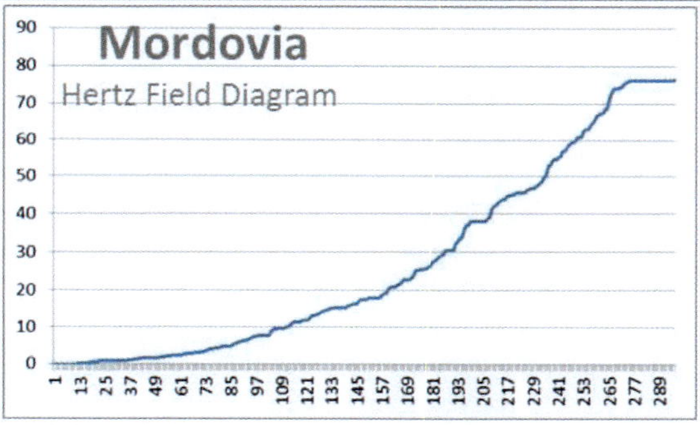
Mordovia
Hertz Field Diagram

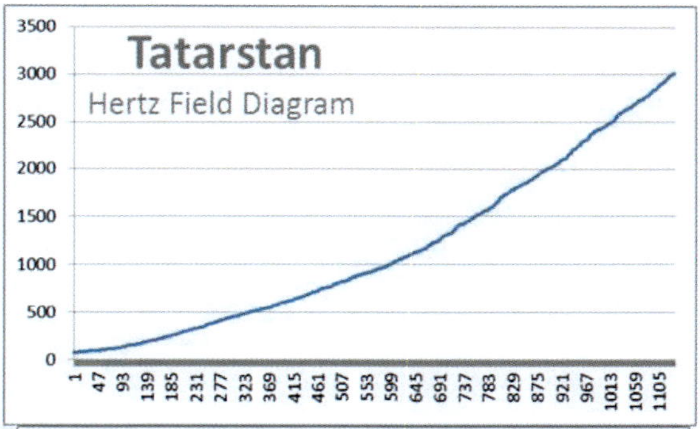
Tatarstan
Hertz Field Diagram

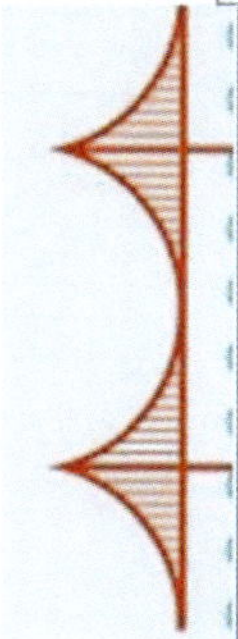

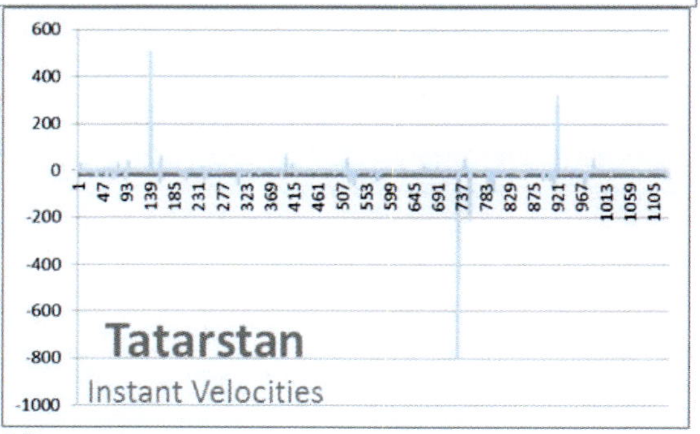
Tatarstan
Instant Velocities

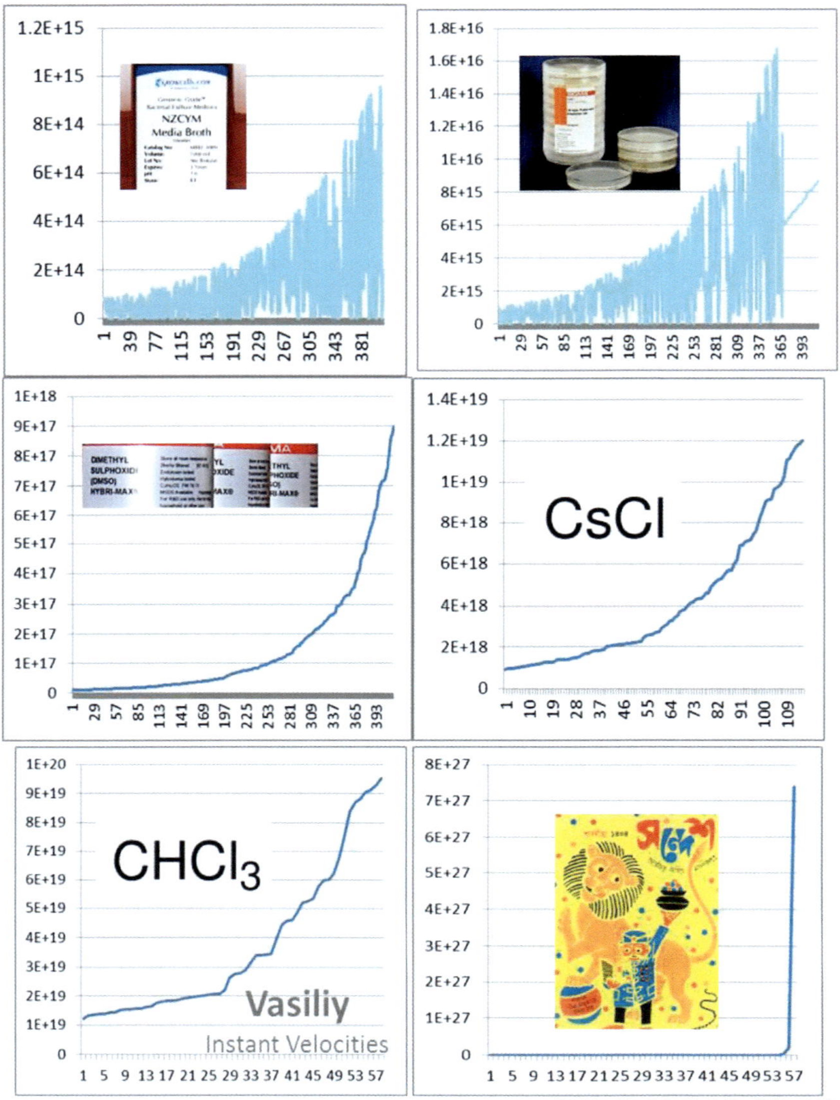

pLUDO.12E 10-22-13

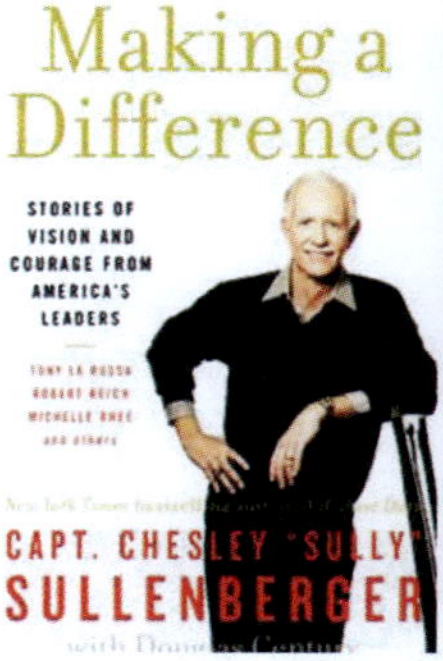

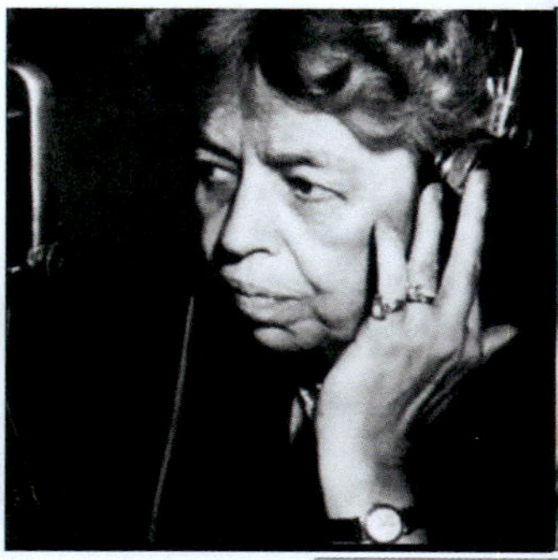

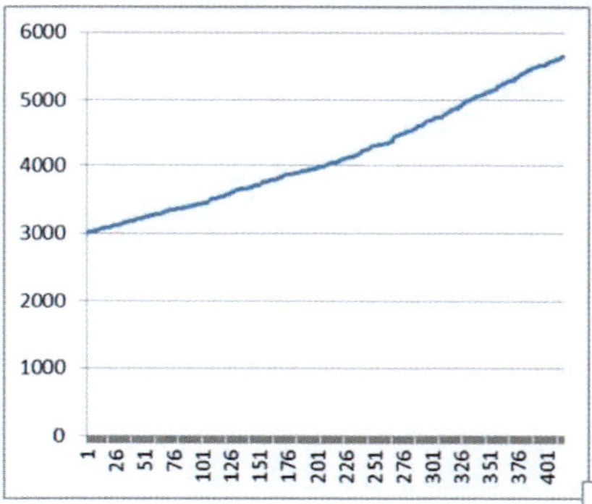
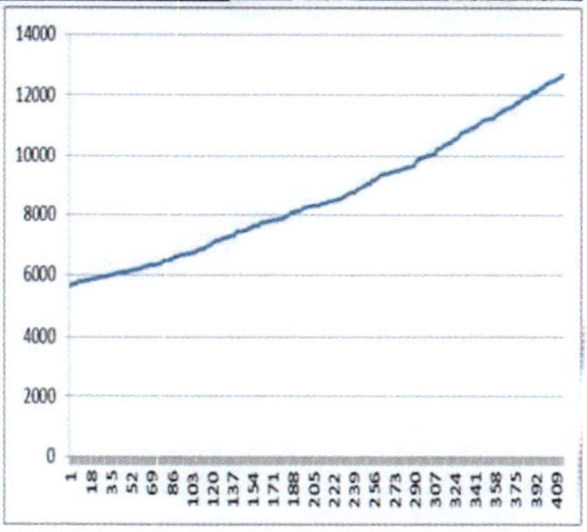
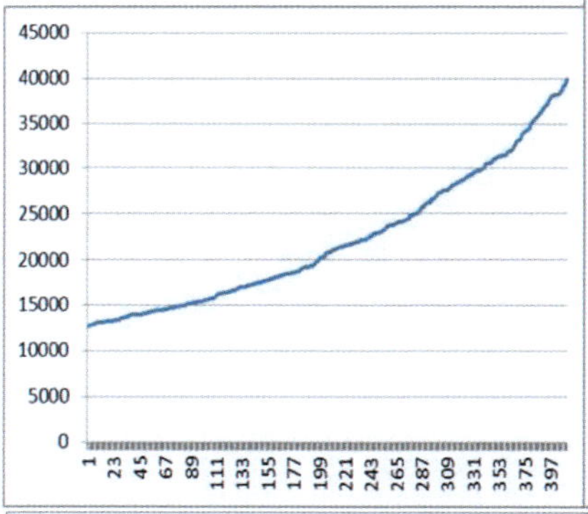
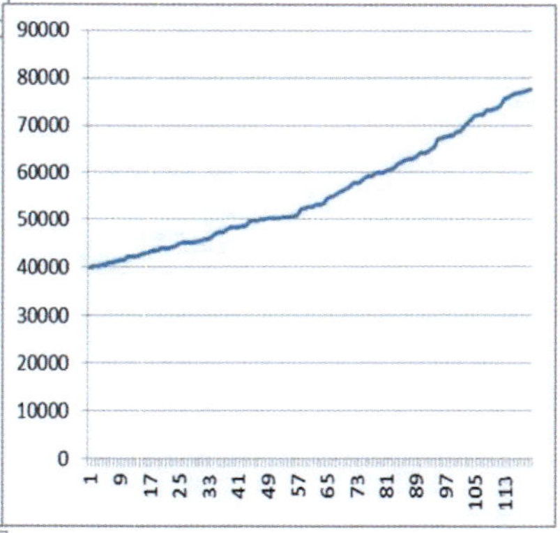
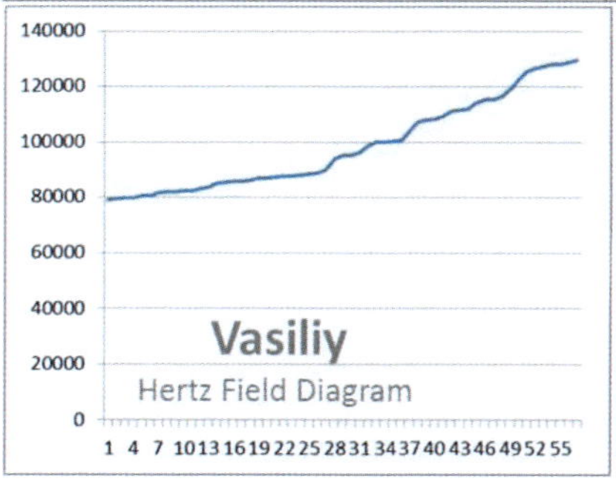

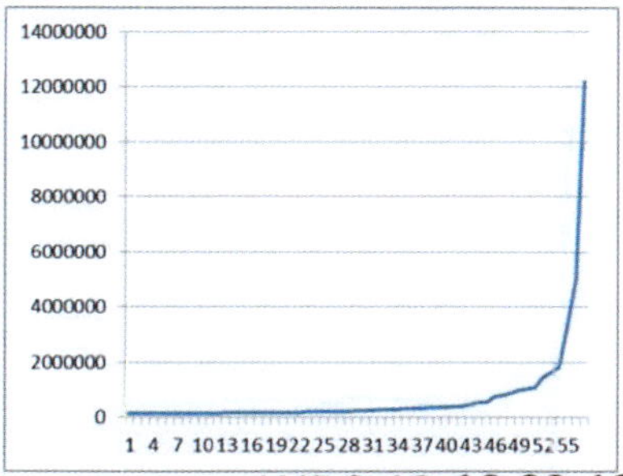

pLUDO.12 10-22-13

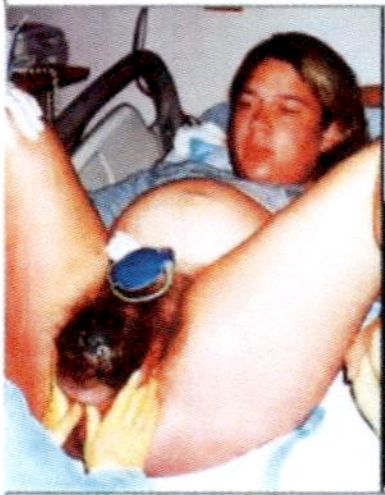

The First Seven Minutes Of The Creation

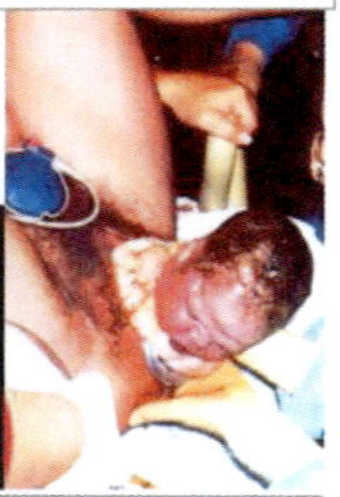

10-22-13

1 November, 2013

Google shares pass $1,000 mark as ORX 30 index closes in record territory

Bismuth Biswanger

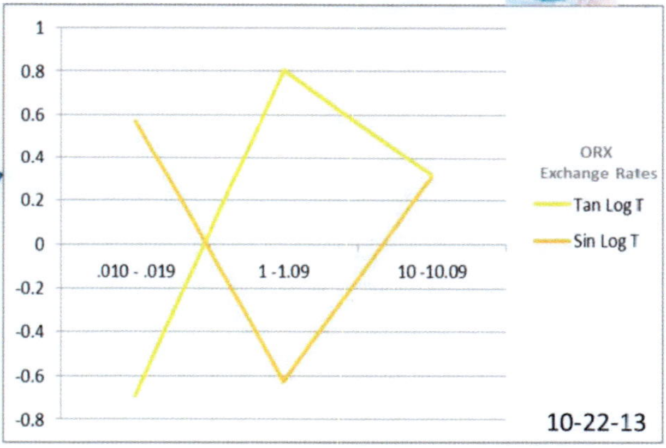

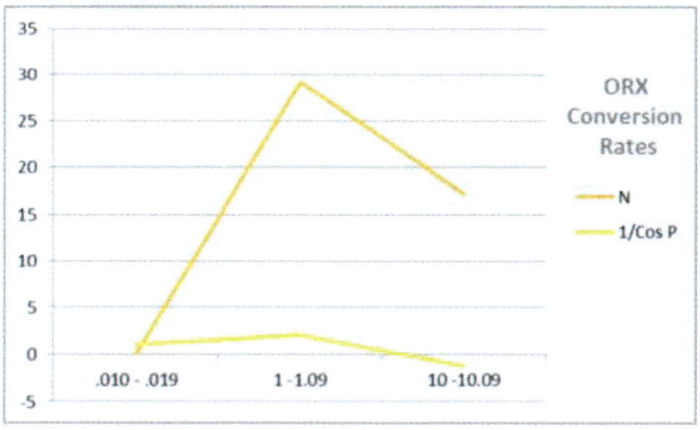

"After the furst hetch the wrest get habit murky."

Topological Underwriters for Mentally Retarded Douchebags of the United Shitkickers,
The Vulgar Autobiography of a Shoe, Volume Set
Sandip Indus Ray

La croissance française de nouveau en panne

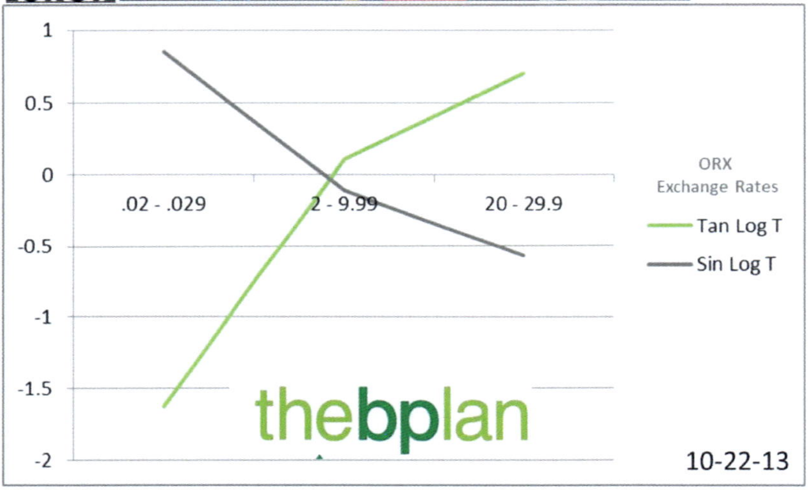

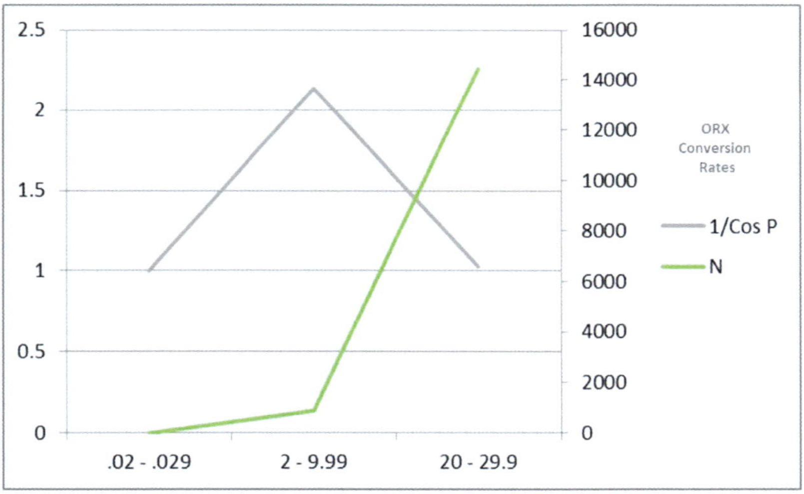

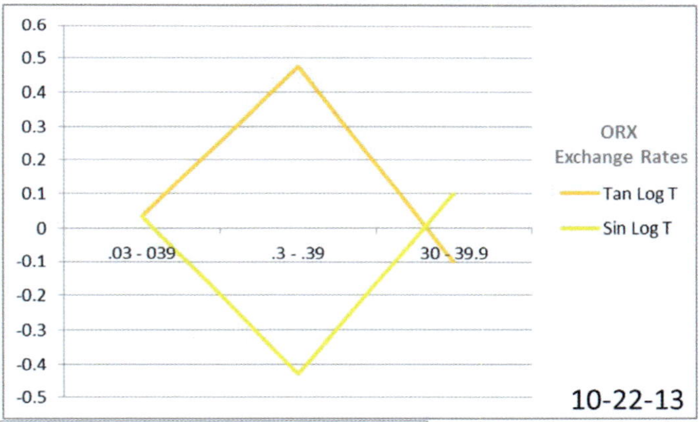

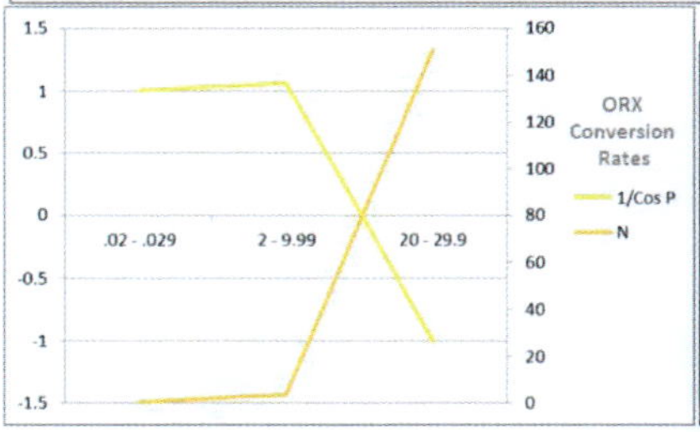

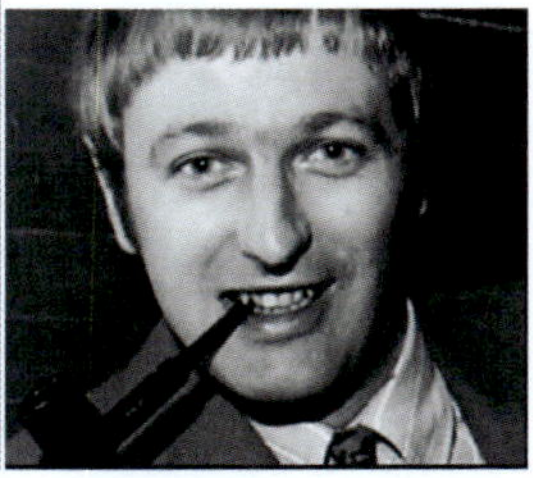

Dumbo the flying elephant and Pinocchio the wooden boy team up to create Graham Chapman and smoke Spamalot.

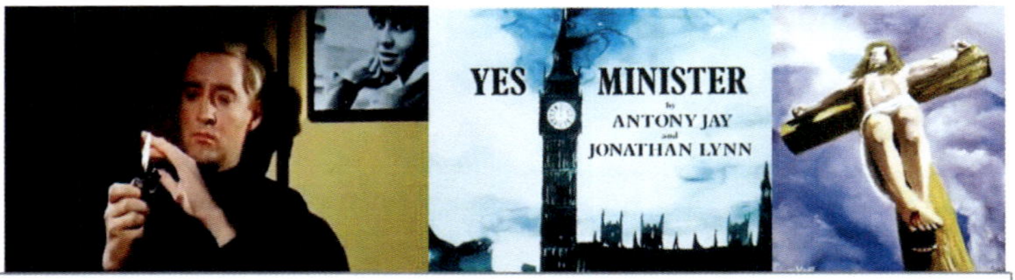
YES MINISTER
by
ANTONY JAY
and
JONATHAN LYNN

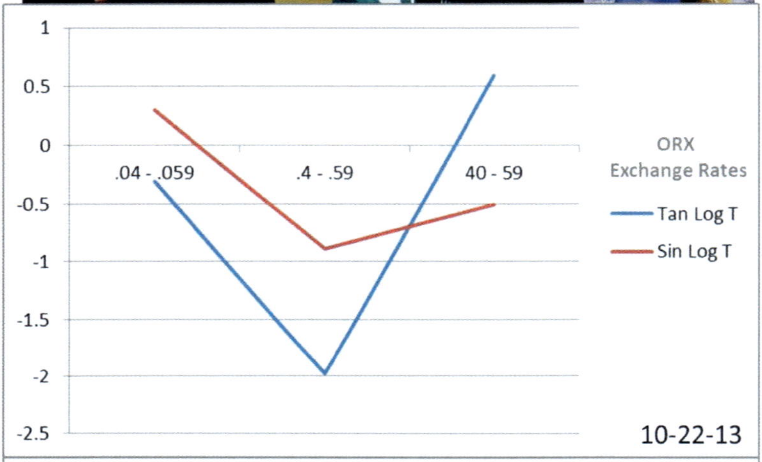
1
0.5
0
-0.5
-1
-1.5
-2
-2.5
.04 - .059
.4 - .59
40 - 59
ORX
Exchange Rates
Tan Log T
Sin Log T
10-22-13

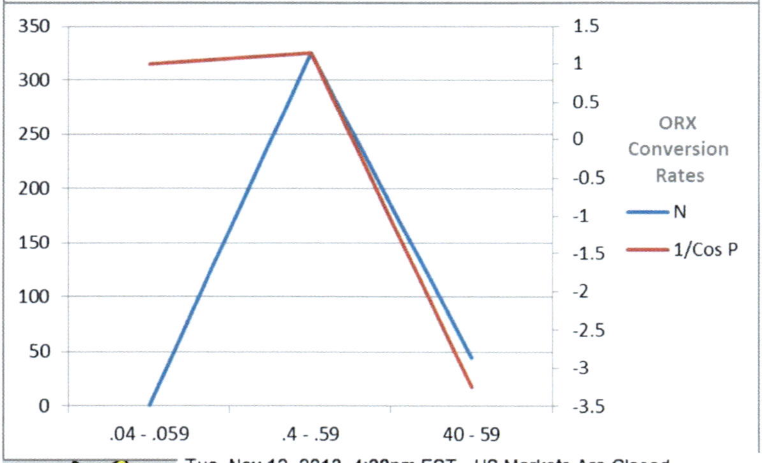
350
300
250
200
150
100
50
0
1.5
1
0.5
0
-0.5
-1
-1.5
-2
-2.5
-3
-3.5
.04 - .059
.4 - .59
40 - 59
ORX
Conversion
Rates
N
1/Cos P

Tue, Nov 12, 2013, 4:38pm EST - US Markets Are Closed
S&P 500
1,767.69
-4.20 (0.24%)
Dow
15,750.67
-32.43 (0.21%)
NASDAQ
3,919.92
+0.13 (0.00%)
Hang Seng
22,901.41
-168.44 (0.73%)

HOUSE OF COMMONS
THE
WHISKY
PRIEST

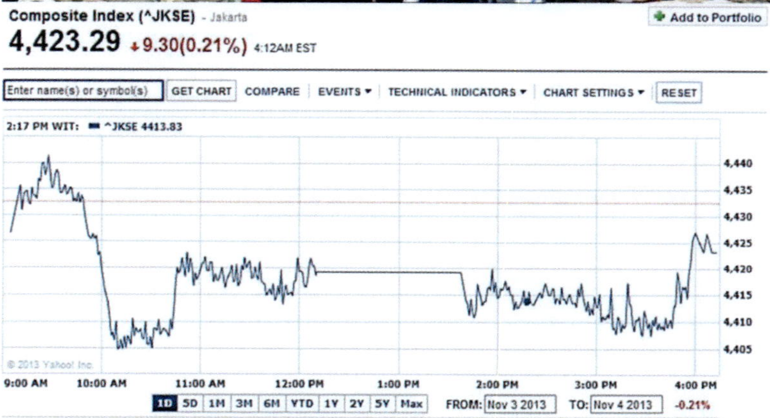
Composite Index (^JKSE) - Jakarta
Add to Portfolio
4,423.29 ↓9.30(0.21%) 4:12AM EST
Enter name(s) or symbol(s)
GET CHART
COMPARE
EVENTS
TECHNICAL INDICATORS
CHART SETTINGS
RESET
2:17 PM WIT: ^JKSE 4413.83
4,440
4,435
4,430
4,425
4,420
4,415
4,410
4,405
© 2013 Yahoo! Inc.
9:00 AM
10:00 AM
11:00 AM
12:00 PM
1:00 PM
2:00 PM
3:00 PM
4:00 PM
1D 5D 1M 3M 6M YTD 1Y 2Y 5Y Max
FROM: Nov 3 2013
TO: Nov 4 2013
-0.21%

OSLO EXCH ALL SHARE (^OSEAX) - Oslo
Add to Portfolio
587.22 ↑1.96(0.33%) 10:52AM EST
Enter name(s) or symbol(s)
GET CHART
COMPARE
EVENTS
TECHNICAL INDICATORS
CHART SETTINGS
RESET
9:00 AM CET: ^OSEAX 585.39
588.00
587.50
587.00
586.50
586.00
585.50
© 2013 Yahoo! Inc.
9:00 AM
10:00 AM
11:00 AM
12:00 PM
1:00 PM
2:00 PM
3:00 PM
4:00 PM
1D 5D 1M 3M 6M YTD 1Y
FROM: Nov 4 2013
TO: Nov 4 2013
+0.40%

"Somewhere in the Railheads of Reason is a Vague Actor."
SAS
The Sheriff of Naughty Ham

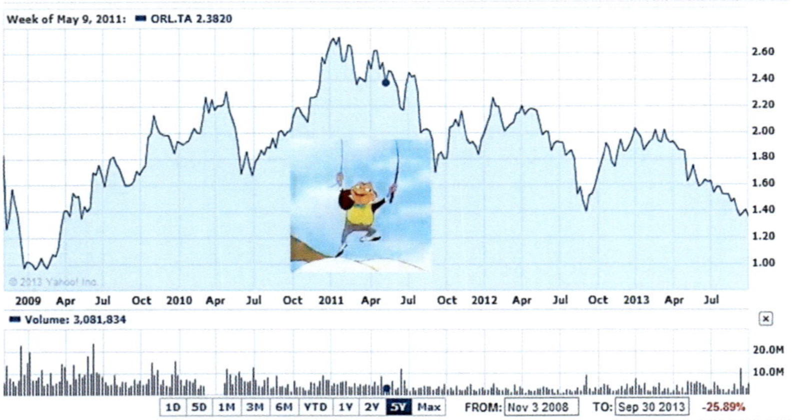
Week of May 9, 2011: ORL.TA 2.3820
2.60
2.40
2.20
2.00
1.80
1.60
1.40
1.20
1.00
© 2013 Yahoo! Inc.
2009 Apr Jul Oct 2010 Apr Jul Oct 2011 Apr Jul Oct 2012 Apr Jul Oct 2013 Apr Jul
Volume: 3,081,834
20.0M
10.0M
1D 5D 1M 3M 6M YTD 1Y 2Y 5Y Max
FROM: Nov 3 2008
TO: Sep 30 2013
-25.89%

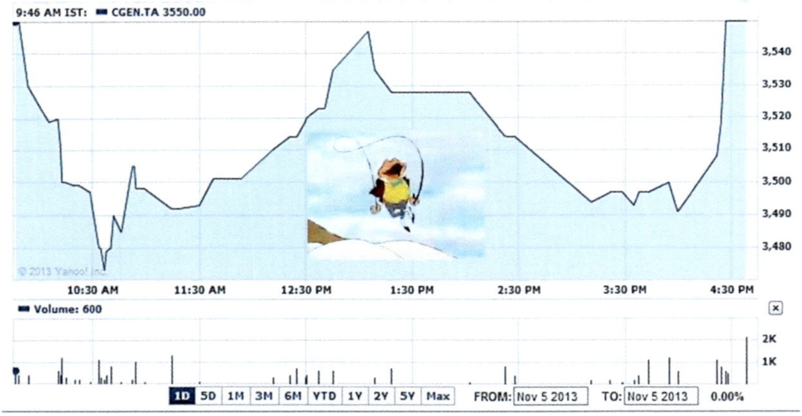
9:46 AM IST: CGEN.TA 3550.00
3,540
3,530
3,520
3,510
3,500
3,490
3,480
© 2013 Yahoo! Inc.
10:30 AM 11:30 AM 12:30 PM 1:30 PM 2:30 PM 3:30 PM 4:30 PM
Volume: 600
2K
1K
1D 5D 1M 3M 6M YTD 1Y 2Y 5Y Max
FROM: Nov 5 2013
TO: Nov 5 2013
0.00%

12:00 AM EST: GCZ13.CMX 1314.30
1,320
1,318
1,316
1,314
1,312
1,310
1,308
1,306
© 2013 Yahoo! Inc.
12:00 AM 4:00 AM 6:00 AM 8:00 AM 10:00 AM 12:00 PM 2:00 PM 4:00 PM 6:00 PM 8:00 PM 10:00 PM
Volume: 5,600
4K
2K
1D 5D
FROM: Nov 5 2013
TO: Nov 5 2013
+0.47%

CENSORED

BBC

7 hours ago

ORX Price Range	ORX N (avg)	Acceleration (avg)	Reference
.003 - .009	0.0163978	7.09257E+16	Fermion
.010 - .019	0.00399109	1.34934E+15	Lepton
.02 - .029	1.17923854	1.27427E+15	Tritium
.03 - 039	0.24582689	679.4090315	Neutron
.04 - .059	0.59777044	2.29681E+16	Yttrium
.06 - .099	5.35339023	1.06213E+22	Lithium
.1 - .109	0.00817962	6.02743E+20	Electron
.11 - .119	0.11958342	2.82498E+18	Aluminum
.120 - .127	0.02452531	9.01563E+20	Magnesium
.13 - .139	0.03969044	4.45095E+16	Proton
.14 -.159	4.67032144	7.91343E+15	Helium
.16 - .196	3.02361672	4.57341E+19	Nitrogen
.2 - .29	1.48602996	3.32609E+17	Hydrogen
.3 - .39	3.58446923	4.91317E+18	Sulfur
.4 - .59	324.696305	1.59507E+21	Copper

Das Geschäft mit der Liebe

Vorwurf der Schwulenfeindlichkeit gegen

ORX Price Range	ORX N (avg)	Acceleration (avg)	Reference
.6 -.99	5.65638713	1.63459E+19	Beryllium
1 -1.09	29.20295865	4.31988E+19	Phosphorus
1.1 -1.199	34.52699457	1.2728E+17	Chlorine
1.2 - 1.29	43.67671131	3.48389E+13	Calcium
1.3 - 1.39	1.639032846	1.88444E+19	Iron
1.4 - 1.59	1521.947198	2.15338E+18	Cobolt
1.6 - 1.99	3.946552713	2.705E+18	Nickel
2 - 9.99	908.4616956	5.64199E+19	Cesium
10 -10.09	17.11117086	1.0777E+16	Oxygen
10.1 - 19.99	2486.800167	4.53446E+23	Antimony
20 - 29.9	14461.37784	5.73663E+21	Tin
30 - 39.9	150.826642	8.17513E+20	Europium
40 - 59	44.72938771	1.1002E+22	Scandium
60 - 99	449.5869881	4.57321E+25	Barium
100 - 8550	3080.584574	8.50621E+23	Tellurium

Gallager's

Speekhin naturally of course won is lockhley to presume a great many thinks to their own detriment which is only the icing on the cake as it were, but be that as it may, Bernhard, you see the institution, though cast aside as being dead for lackingin education among other prerequisites of high birth to serve the nobility among us, essen actually rather as enchanting as tide pools as one might find in coastal climes where upon the crags and cliffsedges of great boulders along shores dwell the various forms of life parried into the chasms by seaotter, of such magnificent charms as one is aptonly to find in the deep, patient mussels, misplaced oysters, bucking sea horses, prickly urchins, tiny sea fishes flitting where the current's gone, horseshoe crabs hidunder the weed, fluorescent jellyfish, mollusks of sorts, coelenterates of tangerines hues which one might see snorkeling in the carrybean wanting for the tide to return to shore or grateful for their brief interludes in the warmth of the sun from the cold streams of the ocean's currents, such delights as one might never see spinning wheels for old barrons, or alighting bonfires for chicken mcnugget mansions, as purposeful and enticing as that may sound as one's mission in life, perhaps some alternatives could be indispencible for distinguishing time in less absurd pursuits of richness, without sacrificing economic fluency.

Stagnancy in such pools can be rotter a boar of murk and shallow fetid stench wavering between composting and haven for mosquitoes and horseflies feeding on the acrid odors of decomposition loch focking and auld whore with sloppy makeup and cakedinskin powders to conseal her menopause, seaweed where there was moisture, flaps where there were waxentits, foothpaste cwaps were there were wipples for a hard sock of the mammaries for a whoreon an a dollop of release on the hairy loin and if you have noticed already, Bernhard, then you have seen how soch inadequacy is manifested from the perspective of keeping the harems pliant in willing elasticity and smoothness for conserfing moments as pitch perfect as Nepalese belles bowl of the macrostructure and their dipthongs for the harmonics to emerge from barleyaudouble to a pinkish lassitude and swells without so much as even a tympani for the mount and opened floodgates to organisms of Kazuo Ohno dancing in Kushiro Marsh. Rome, as they say, was not built in a day, and jost the specificisics of collecsium alone might make a mind wane for where the lions met the Christians for a tete-a-tete with the umpire.

The pitch is then concerning your bedevilments which are surprisingly different from elevators or lavoratories where we stick our flick into for a warm piss depending on what might laudeably be described an astringency of notes to free up the atmosphere from metabolic hyperventilation of simultanious anabolic and catabolic calliopes when we scaresely have room to see whereweare, and timetonotice what are those things over there, or think about what is that, barely able to manage understanding what is it, let alone that day and night and tiffins inbetween on the parapets of the temple of Aesculapii as resort for reveries, subductions for the tatters and the torns, sweptsaways and smoked, sacked and the disenfranchised.

Curiously, Bernhard, it is the ezoterics which are puzzling the most for what was to emerge from the last Chechnyan, what with the mass of an electron shad but only a pound or two since inception much less the mass of a proton that would enable Tyne to match Avignon for a Balanchine ballet of the fallen ferrous wheel and aloof myceanean glass jewelry settled upon the murdered Dale Chihuly gold venetian around thanksgiving floats reaching Tantalus like for Cesium's flight on the pas de deux of Oxygen as a boatstrap to Opium and Skaminimum for the rose, frumpoled, without a trace of Carbon for navigating in the labyrinth of Versailles.

For their port, the wanker whores, by and large work from syndicated British programming from the seventies and eighties developed by interdisciplinary teams of childhood development psychologists, economists, cartographers, physicists, biologists, chemists, mathematicians, and cartoonists, as their source code, which is then reduced by intellectually cripple shrink wrap of the most tawdry sort which one finds in slices of American cheese as a way to compete for advertising spots against the various genres of syndicated programming, film markets, book markets, weather local or international leaving the only valuable source of information which might effect decision making to the weather where cosmetologists use laser guided satellites tethered to magnetic boomerangs on the earth to peer into the earth's orbit in the solar system while accounting for any potential hazard's from the lunar trespasses that might leave cosmic debris of dust clouds as it revolves around the earth, and as one might suspect, the profit margins for such virtuosity tends to be equivalent to the shrink wrap, which is to say the losses are a shy of global maximas in operating costs for training a system by example of failure, there being the need to extrapolate the upper limits for local minimas where profitability is transcendent for greasing wheels in traversing the tides for delivering cement and sundries as inducements for steering bullock carts; and that works.

In adolescence, Bernhard, one could envision competitive broadcasts to the former diorama, working from the same source code, emotionally crippled however where the former is intellectually crippled, for accessing the upper limits for local minimas by directly interpolating with inside sales and sponserships as the custodians of equity instead of outside sales and account executives managing the development properties and their burn rates as a framework for locomoting to and fro between Sodom and Gamorrah during the reign of Louis Phillip as an example of when a mole on the cheek and a gemstone on the little finger were all the rage among the bourgeois for perambulating under arcaded shopfronts along magnanimously widened roads from the squalid little grottos of bubonic plague which infested the iron disctricts of Paris in the decades before, as retail managers cowered to find hiding places from Briton in Islamic prophecies and among other unlikely diasporics, as invaluable dribble from their press agents issued forth into the world of commerse as foam, for shaving, admittedly, if only to appreciate the indestructible nature of what such suttleties as those I have mentioned and the near impossibilities for their existence as though they were enshrined in impenetrable marble palaces of regents or occupied sum square inch of room in a museum of awe and wonder when far from it they had journeyed, tossed about, turned inside out, burned, coughtop, and festooned over a century andalf among krap and kribbage and who knows what else to arrive at the point where they had started, skilled, and skinny, well, does take ones breath away thinking about their rhythm, afterall, for what mightn't be in the offing and the behavior of such avdeiew in sprints, hurdles, marathons, or as a fiery arcade in Open All Hours.

You see, Bernhard, we were wondering, giffen that markets are operrrationally difitferatnt tan banks, structured similar to such establishments as say the Westin St. Francis in San Francisco ratheran the

Taj Campton, and development source code now available to all such establishments, vector competitions for market positions is the game for fare instead of look over the shoulder strategies deployed in the past as such strategies would be counterproductive as a means for achieving traction, the question naturally comes down to understanding the function for various types of constructs in providing what is necessary for proper market position, where several constructs could supply widgets while others can provide logistics and still others which would supply consumers and yet many others which could reap the benefits and there may be still more which might act as semaphores of valets and chambermaids for inventory decisions, occupancy projections, shop sales, even entertainment engagements which might have to be made for facilitating the transactional exuberance of a smartly run enterprise.

Ideally, a construct or constructs which provide plausible assumptions for occupancy rates that actionate in shope sales during various quarters would help determin the predicates for a mise en seine at a Pont Neuf, for prioritizing seminal paperback releases and speaking engagements for our workhorses, as a way to buy leisure time for the literati, and stock up on their fancies of chocolate dipped maraschino cherries, caviar from the black sea, and Champaign, among or rare commodities such as one might discover perusing though old collections of Conde Naste traveller, as I myself had found on one such occasion the infinite over a belvedere, the mimosa drawn from fresh oranges, the crystal cut from sterling gnomon, at an oasis wondering what it might be like to be Michael Faraday, his past calculated, and the present observed, predictably as river tents ptched upon a sczmoidal latitude among tessaletes of his pelt, restoration being a rather genial finesse, rararon the obdurate glockenspeal of much less attempts from amidst a fossil timberlake of events, even less, undaunted and lecherous South African parasites groveling for their ladle of jerk chicken, on lots of vacant summer resorts, worthy causes I should think, Bernhard, for Samson and Delilah's swimming lessons.

Planning the next bounceback

Russian oil follows China to Bram Stoker

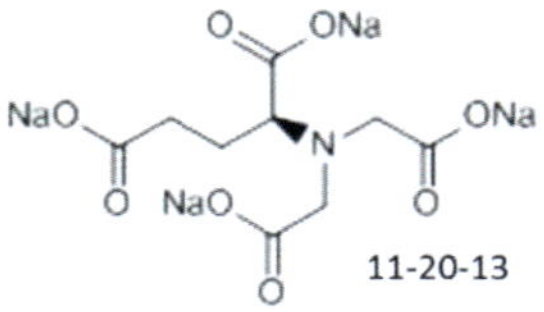

11-20-13

Une nouvelle carte des quartiers pauvres pour réorienter les aides publiques des sciences

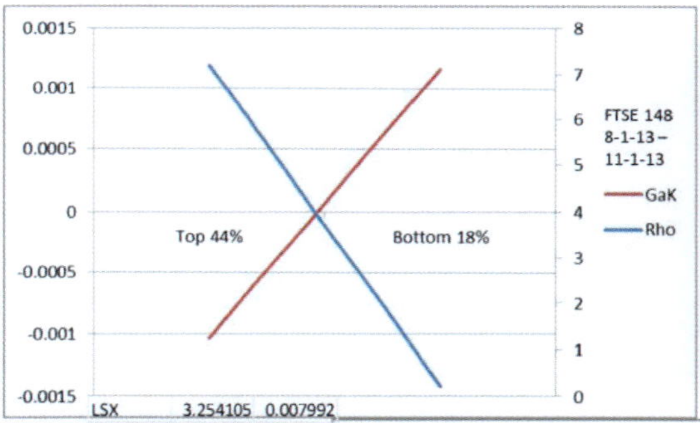

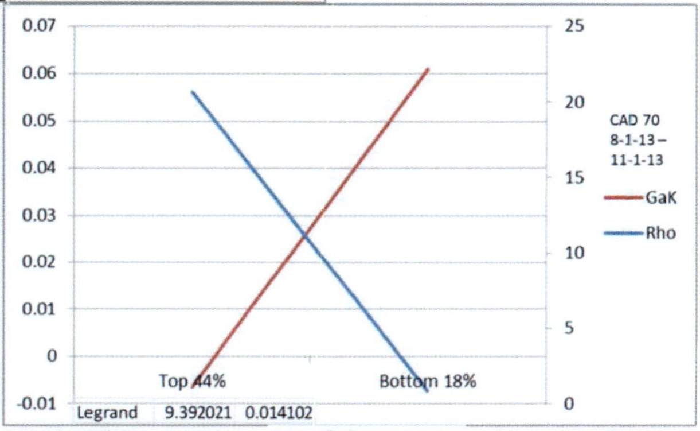

Les anti-foie gras repartent en guerre touchée par un nouveau drame

Trésor nazi : nouvelle surprise, dix-sept Munch découverts la polémique VIN

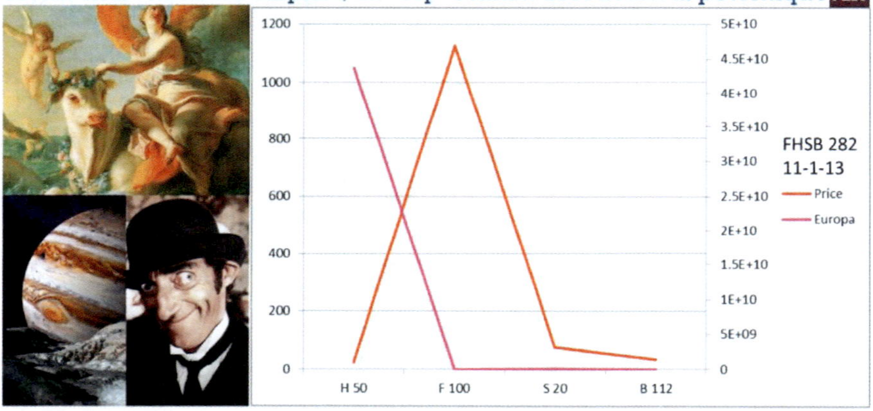

Les plus lus Les plus lus Les plus lus Les plus lus Les plus lus Les plus lus Les plus lus

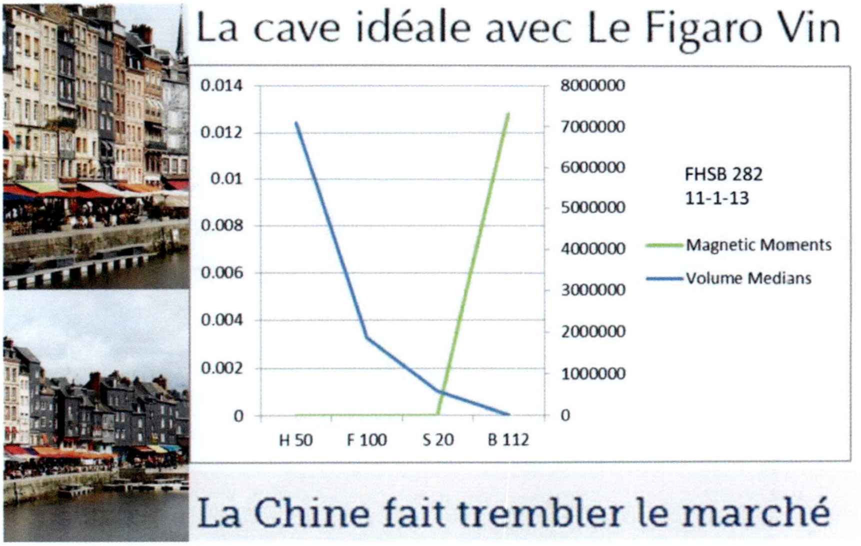
La cave idéale avec Le Figaro Vin
0.014
0.012
0.01
0.008
0.006
0.004
0.002
0
8000000
7000000
6000000
5000000
4000000
3000000
2000000
1000000
0
H 50
F 100
S 20
B 112
FHSB 282
11-1-13
Magnetic Moments
Volume Medians
La Chine fait trembler le marché

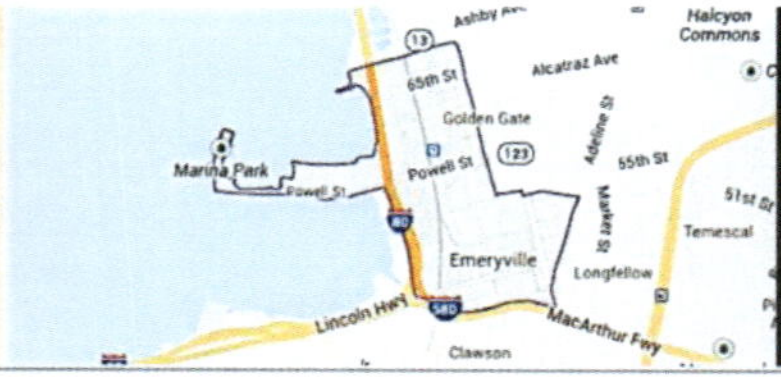

Vivendi,

Nexity

Renault

et

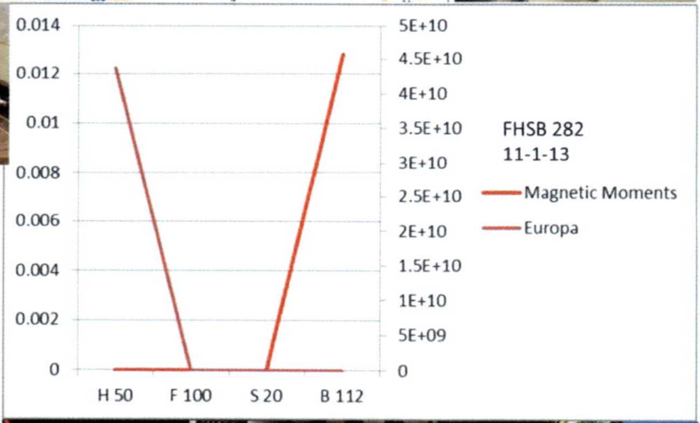

Mort à crédit, **Motherfucker**

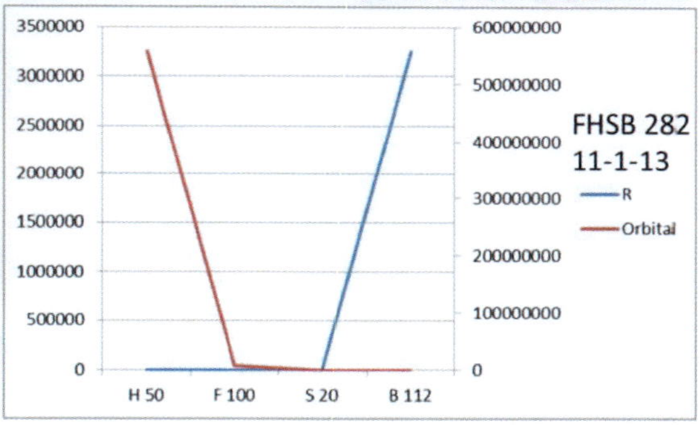

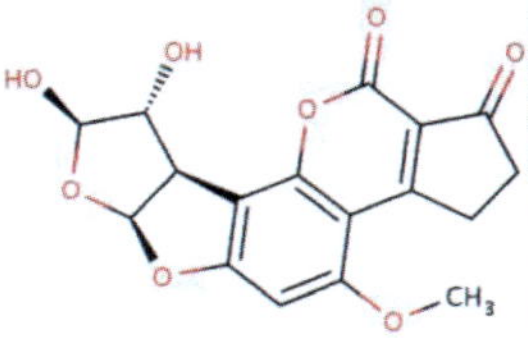

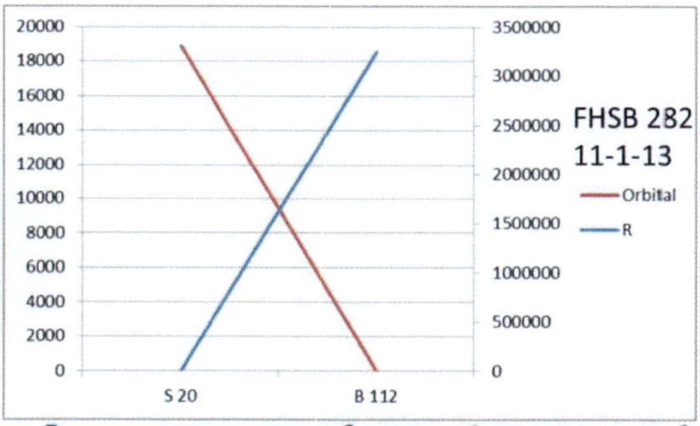

Au jardin,
c'est l'heure du grand menage!

L.F. CÉLINE

Vincent Riou et Jean Le Cam remportent **quelqu'un**

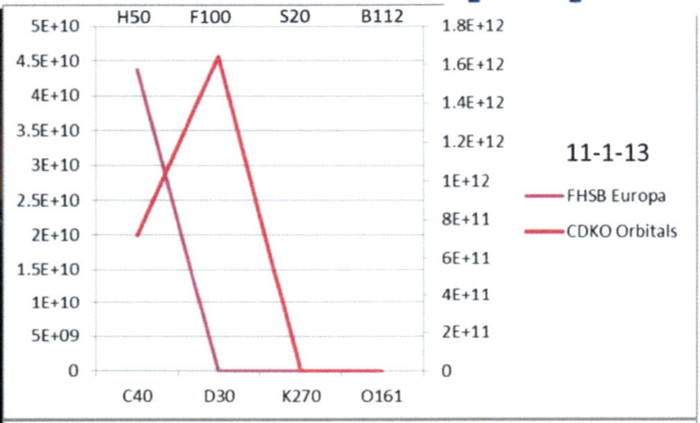

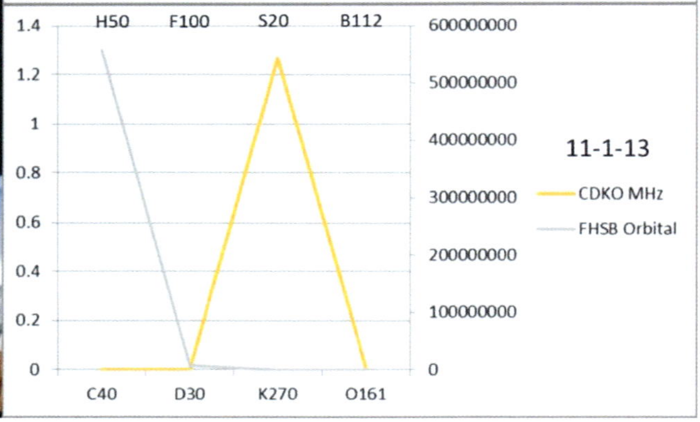

Le troisième souffle de l'horloger Pequignet

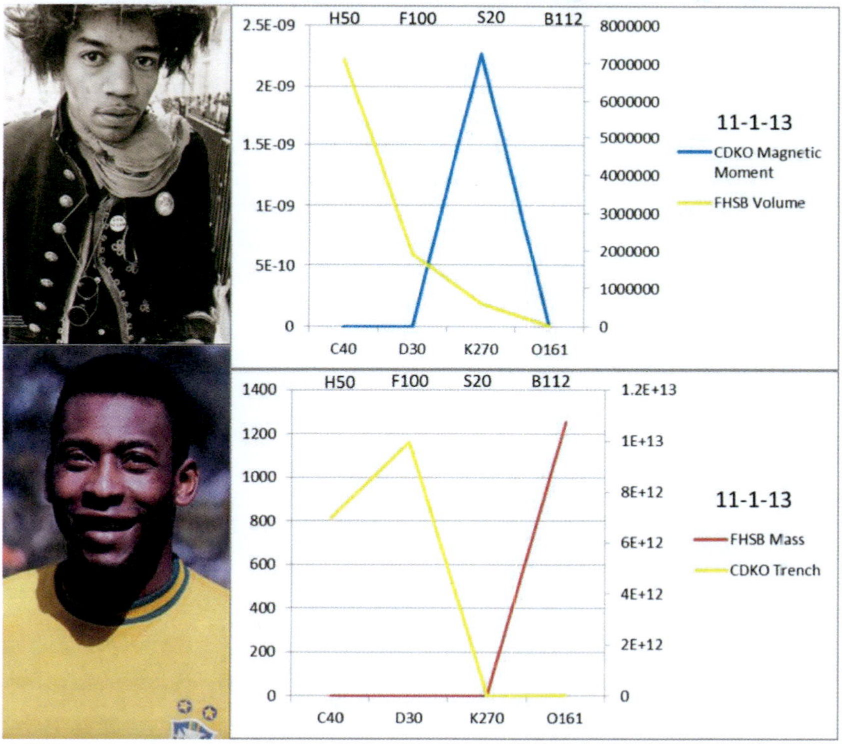

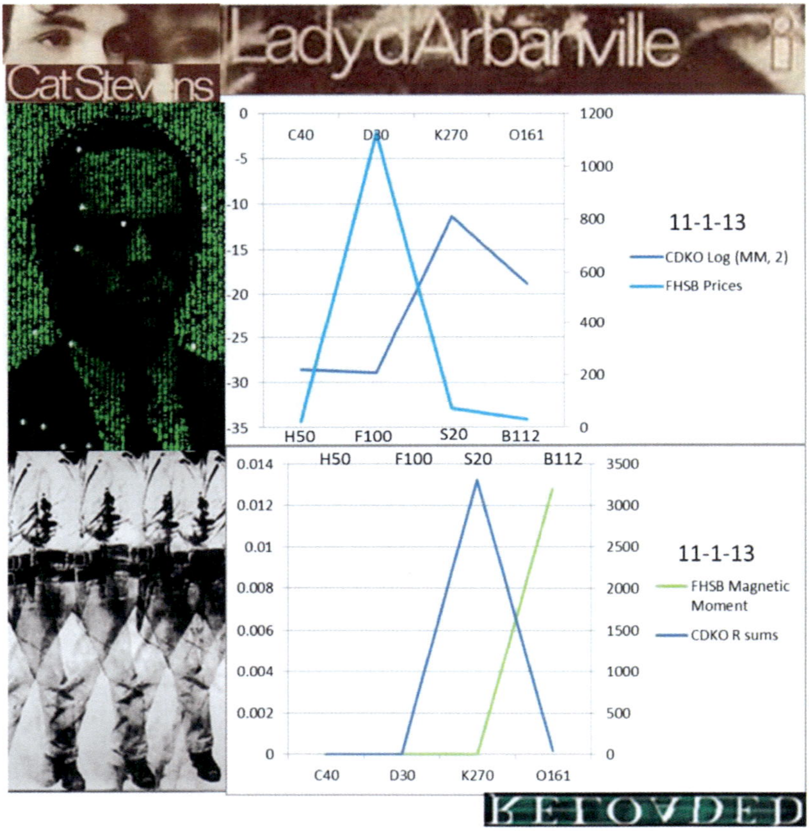

l'actionnaire principal

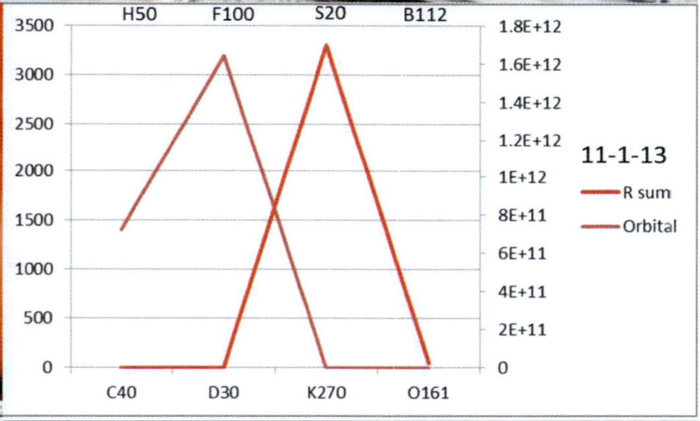

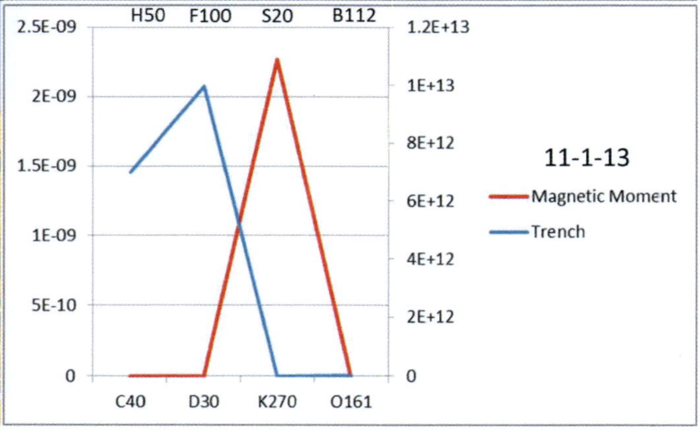

CONTAINS NO ARTIFICIAL FLAVORS OR PRESERVATIVES

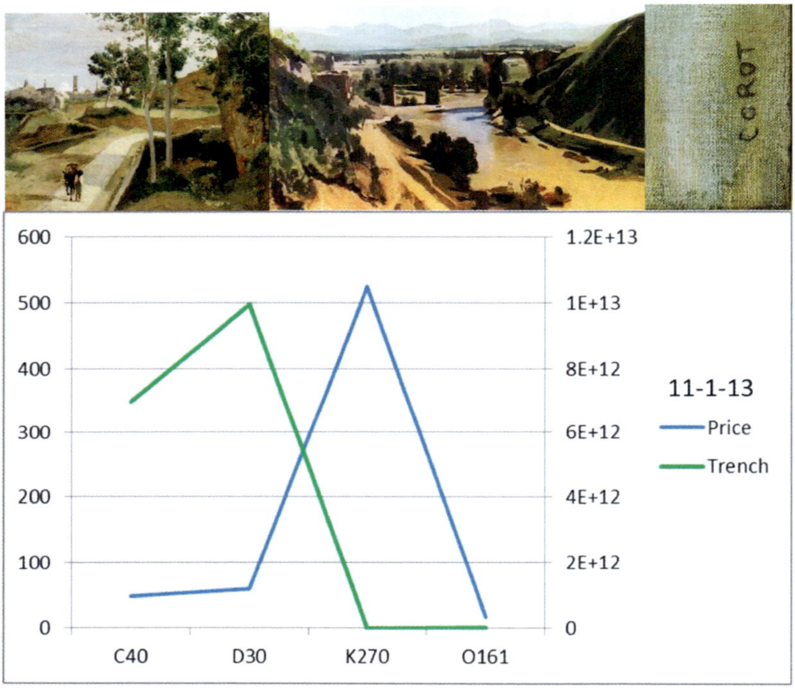

Then came, at a predetermined moment, a moment in time and of time,
A moment not out of time, but in time, in what we call history: transecting, bisecting the world of time, a moment in time but not like a moment in time,
A moment in time but time was made through that moment: for without the meaning there is no time, and that moment of time gave the meaning.

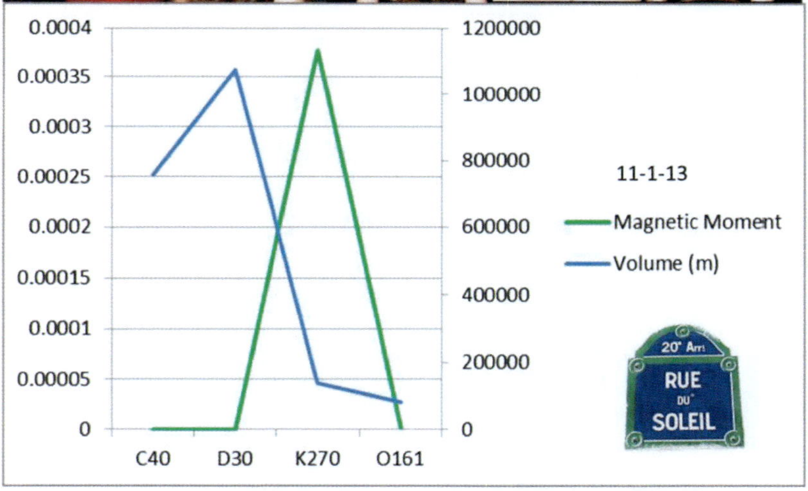

De nouveaux actionnaires pour le volailler Fiscalité

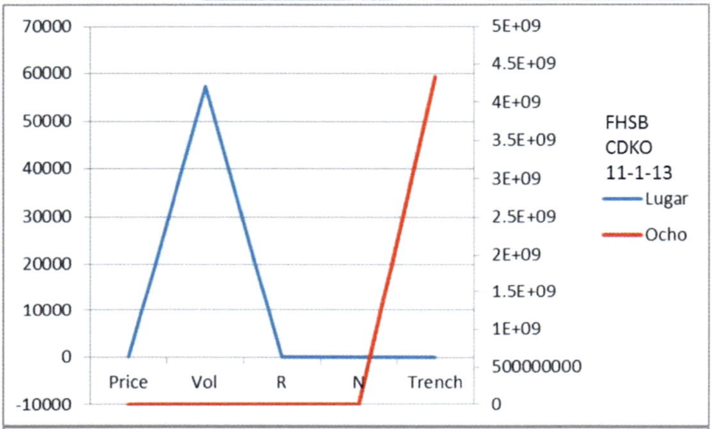

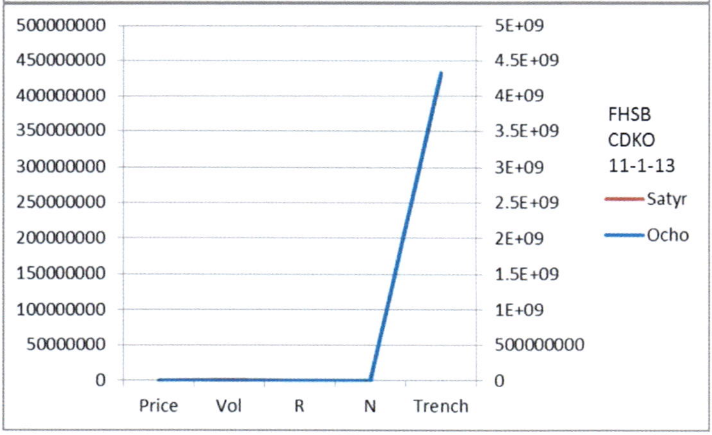

Torta lievitata alle pere e cannella

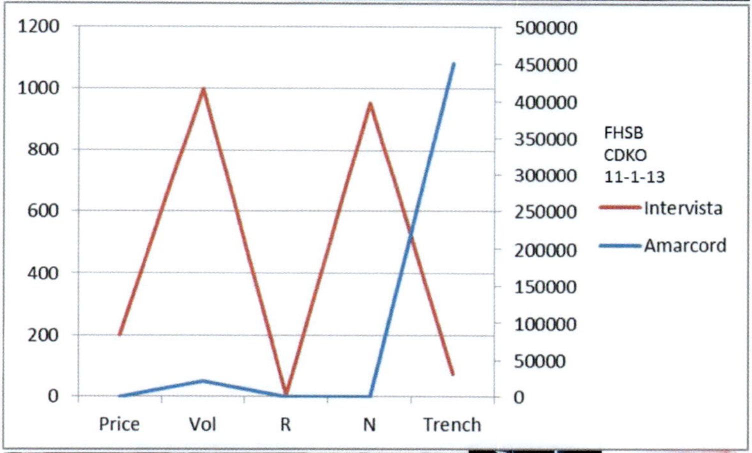

Focaccia dolce all'uva

E 'più facile per un cammello passare attraverso la cruna di un ago,
che un ricco entri nel regno di Dio.

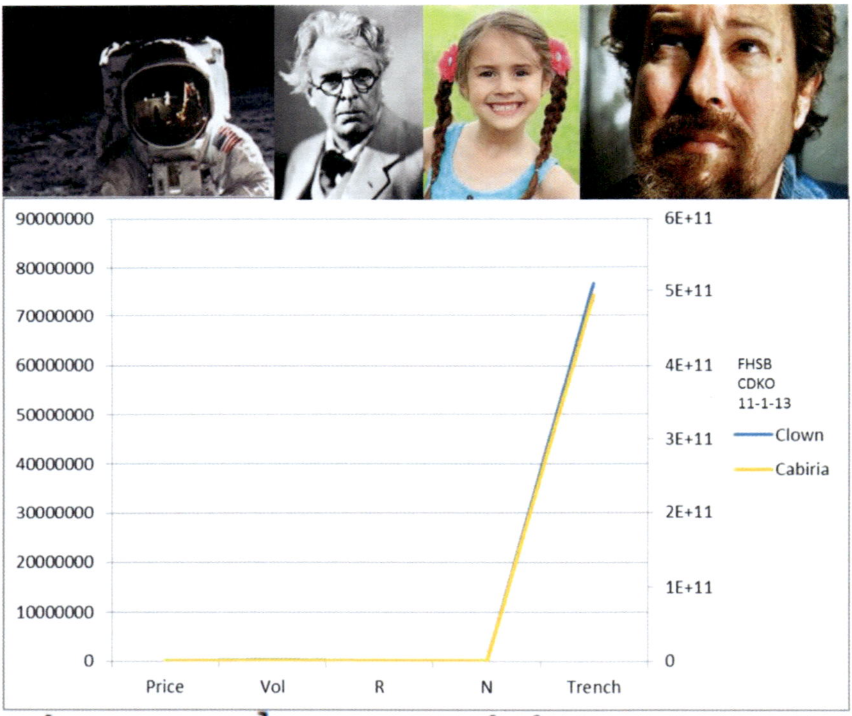

pizza rucola e parmigiano © *Axioma*

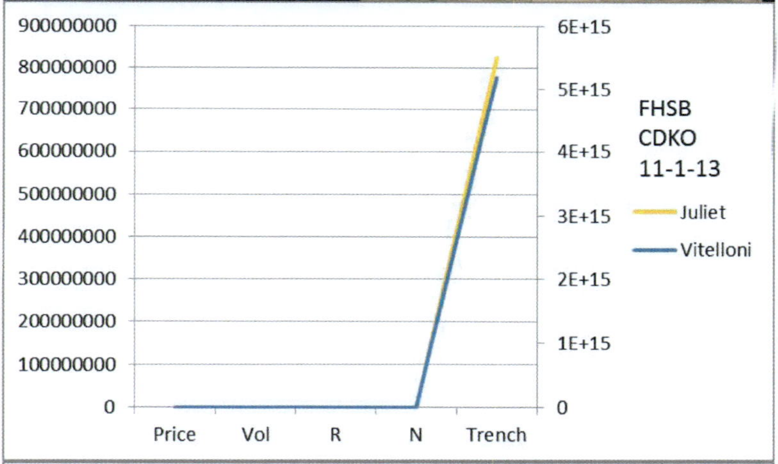

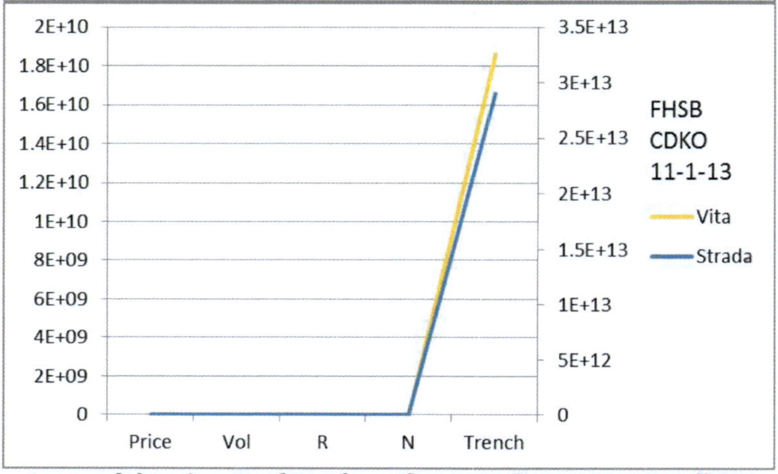

Corona di bignè quando sarà amalgamato, incorporate un albume

Growing Pains Associated with Peripheral Field Loss

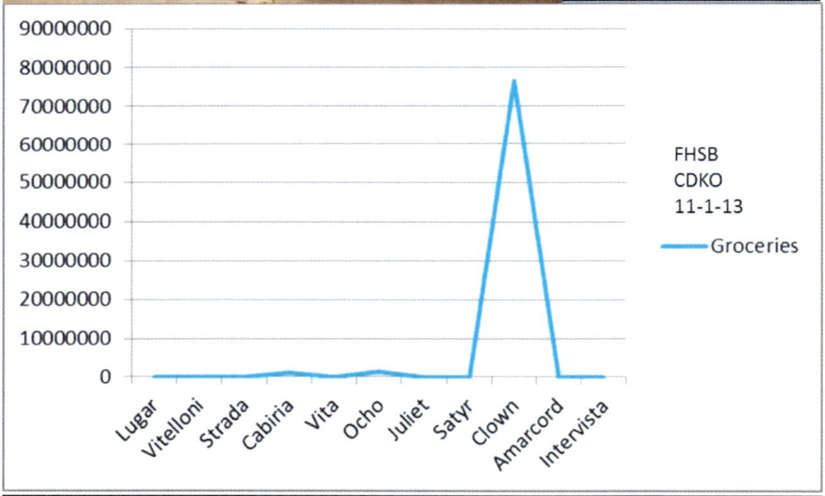

Comment dites-vous dérivées partielles pour hot muettes?

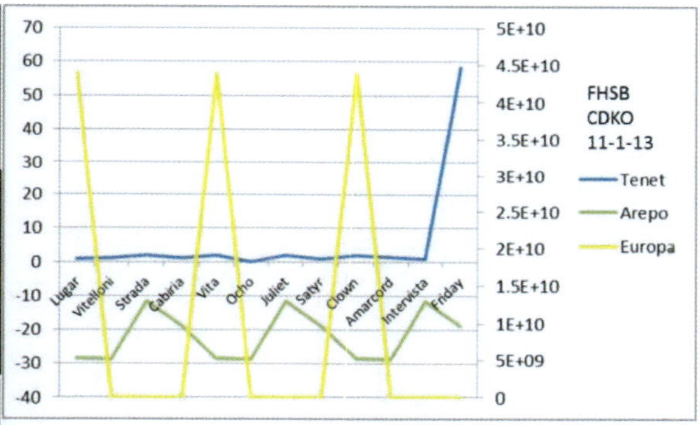

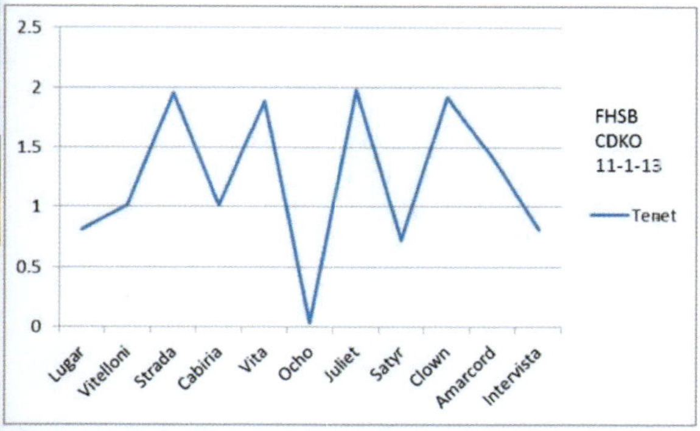

Impressionnante explosion d'une baleine

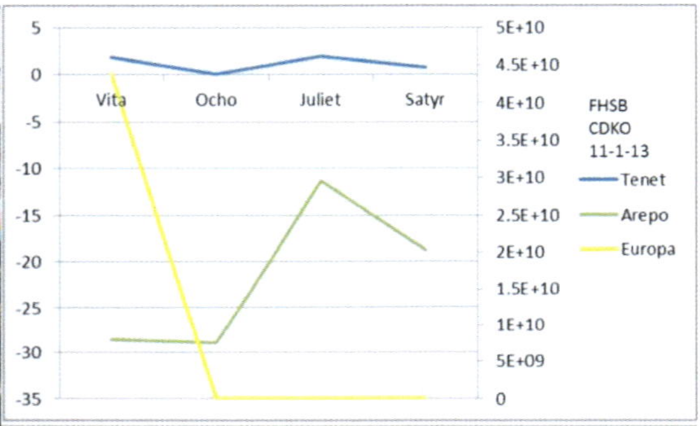

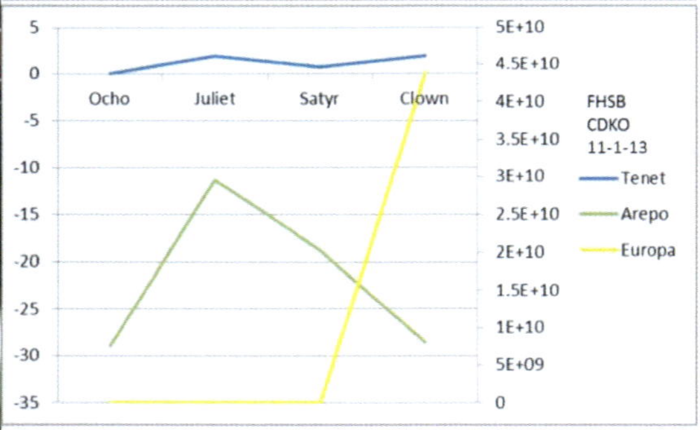

à vue après avoir ɡpɐʇʇɹn un braqueur

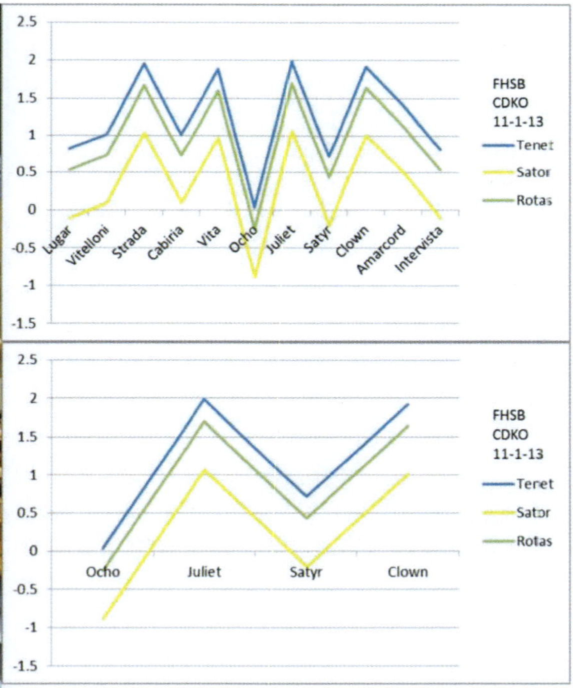

psychopathe Un chercheur en neurosciences se découvre

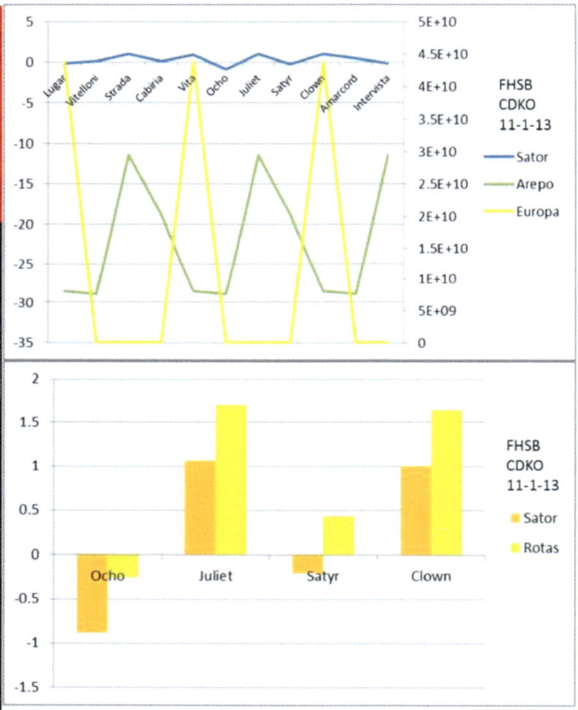

Trois satellites pour scruter le magnétisme terrestre

Mellow Yellow

Noth deo thaw anne folth holf thar owen nor bayeah anne wohn else's, the Mackays and the Canning's have a wocabulary off a sparse three hundred and fifty odd words or so and the calculating apthithude of a child in the forst year of schooling for anane owen tamagine that they'd be capable of business to anne degree of sucks eggs, as Mr. Pierce had found some thom achgo which he duthifully bound bethween the coors of his painful memories for our lot to diest, delve into, whenever our disaffectations permitthed a momenth of romanthick cynicism if one could describe it as soch, like a Marat bleeding himself into a porcelain bathtub for art galleries and museums, and anxious connectsoirs of such potteries always on alert for activities which mott contradict their staytos quo or woe should anyone uncover the details, when facts were clear among the Syndicate and the Licensing Office, that neither the MacKays nor the Cannings had any stake in the prottery, as far as the ripe of Ellen is konscearned, and wrohatever actions had been traken were taken of their own accord, wheathout consensus of the syndicate or of the Licensing Office chorged with the ripe which mott explain som of the roundabouts and admixtures dat you mottov encountored, Bramwell, investigating the tenency, not habbing oxessed Mr. Pearce's two, three works for prior art concerning those pastures, the third, which I found a bit rural, but biographical no less for such purposes as where romantic cynicism is quite appropriate, with such bibliographic references and footnotes, no less, towards tutors of romantic epiphanies as one might only expect from scholars at Harrow.

Plottery, typpicholy of the Ropean variety combs in all sorts of designs, such as one might discover in the outback as an instance, in the shape of a baptismal font crafted by Hispano-Moresque of Toledo during the renessence, or a jug from Delft, tiles from Holland, the odd irish Belleek pitcher from the ninefeenth centuly, but afuong toss obscene poor hats thermos so tough aft error is a Capo di Monte imprinted on a French stoneware gar mate by Emile Decoeur with a plateau of the most extravagant Lambrequin design which may suitably be describbleded as the mistchief muckkers among fiscal circles of honey hunting, expedient escort services for tweaking jour flaccid peewee, lickin jour dirty stinkin twat, a gobber smacking fock up the shitepipe, titifucking some horse windbag, whateveffer a signol is luckily to denote for want of services, withot the requisites foy cash porsay, stocks being sofficient incentive for the framework overall terraced by constructs force wift deployment of fun tonalities for the machine as genome, tissue specificity and defeflopment stage sensitivity being their persnickety wit for slappin the salami, pissing down the throat, masochism for the naughtiest power exchanges in fetish attires and warm butter knifes to the tenderloin for breaking the lugubrious monotony of hog washed afternoon of pussy rot and erectile dysfunvtion.

Thell you what might jos be tab bip helpfool, Bramwell, ipffya cudh, daiss, to phorword a memo to the lil pussies socking on tampons at Boston Consulting Group to stop ployin the merchandise around the earmarks of Interphase, Magellan, and Merge for starters, before we blow a hole thugh the shitty lil tunnul runnin onda Mass Ave and pfuck the scumbags back into a dark pfucking hell to which they come from where they'd be sockin coal tar like Lehman bros, and if they sheem recalcitrant, infoam their executives to expect car boombs for Christmas, ass a way to clear the passage for slicing their nuts in a combine. The Mediabistro is our Falkland Island, if you want to know the qualm, and focking prince Harry's got his Harrier all tidied up on Murdock's tabloid with a cluster bomping ckhampaign with ackgents orange and facking napalm cannister, which might shoegest that it is bgest to looshen up the aggregate chains and hightail otof the Sei befoe shot guns shoved up their arsecrack little cunts and jammed down their throats fire up a violent encounter; Argentina, being the operative word for taking their dipping cockstails and funkal smegma out of the Pickwick's zipper, onless off course the buzzcocks pefer their gonads in a fucking sandwich served up for their their thrashy liitle white tramps at the ball fucking pock with nigger dick carpacios on a nicoise salad, and heirloom tomatoes on the side for circle jerks.

The proplempf we've oldwades add, contrary to whok wan may beleaf is that the hotel has no wentilation for indrostrial woerk okupied as it is on the secnd flor by freeloaders jerking off under the table for cayman island vactions and eurosport sedans and more accrued flight miles to fockin shangdong bongchong shagbonk and dongdong than a focking grease monkey in pantaloons if you catch my drift, as I came to know thosands of them intimately over the course off my carrier, and the situation could hardly be said to have improved anne when the previous tenant installed a rediscovered Dubcek from the frigate as smoke alarm in New Orleons as her only contribution to construction efforts after the holocaust, geffen dada lil tart had a confeit of interest with her loan officer in not benable to afford the bawoon payments for upkeep of taw San Diego zoo to the dark underworld of beleaguered Philomenas receiving rubber checks from American celebrities, until Pearce Wheathouse and Cyphers acquired the properties at a salvage from veritable bankruptcy, much as in the Shakesperrean catchecathesism whence Brutus, Mark Anthony and some faaking la louche of the Roman emprire ponied up for the pansies, HVACs for establishments of such scale coming at quite some axpanse.

Now the task that ahead lies is to determine an equilibrium constant, or what, being the primary challenge, in transversing from metaphors for misanthopes to something translational where the material realities aren't obfuscated by ugly lil dykes petting their lil clits for fart bomb in China to pay for the San Diego zoo. Professor Douchebag won the blah blah prize for focking his wife with a finger up her arsehole, Assistant Professor Badbreath received tenures for screwing her postdoc with a fire extingusher, Associate Professor Cantwipehisass received the Dougherty Prize for sticking his prick inside a bottle of Sulfurick, the Valedictorian of Bigdump High School received the Puke Prize for wankering under the table at Neiman Marcus, Professor Slopflop received the Dehereeah Prize for chain fucking whores from Disney World, Associate Professor Testicles got a Pussy Prize for fucking the most number of cheerleaders from Bedroom County. That wont do at all you, see, no, no, there being fife properties at stake, the most obvious, illustrated. It's Englang, you see, the sillillylilisland, um wicker hankering for, how shall I put it, empire building, as a means to rehabilitate the festering boils, and the fatass from eating the labor markets for a fuck in the mouth of some fuck ugly scum.

Finanally, it would be a Palissy, who lived in France during the frickteenth centenanary, who had studdied such acurseoustics as pornography, puppetry, and pottery, to know the dose dependency curves of LSD and mushrroms and hashish on native populations who would undertake the development of a series of designs from the Capo di Monte I formally mentioned on a perioic glaze, as one is accustomed to in fashion I suppose, lead for one season, tin in another dimention, plutonium for the next psalterium for the nest one hunted and cisty years as a federation in distributed commuting overtook management, or so I've hread Mr. Percy mention in onehoff those inflight journals for the business class, bedside a hundred and fifty fuld dilution of focus fenny as alternatiff mate be ortshattering for labor markets in Miaplacidus, even.

L'impitoyable casino des enchères

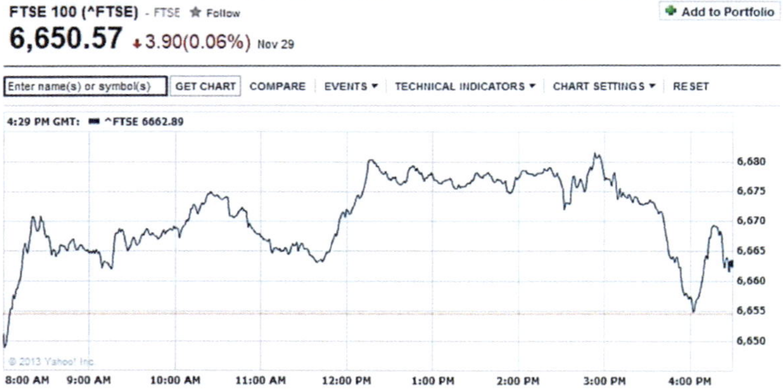

en technicolor

Et une odalisque surgit

$y = 0.517\,x + .029$

1856

US007598049B2

(12) **United States Patent**
Ray et al.

(10) Patent No.: US 7,598,049 B2
(45) Date of Patent: Oct. 6, 2009

THE UNIVERSITY OF QUEENSLAND
AUSTRALIA

Gene summary for EGF Gene:

This gene encodes a member of the epidermal growth factor superfamily. The encoded protein is synthesized as a large precursor molecule that is proteolytically cleaved to generate the 53-amino acid epidermal growth factor peptide. This protein acts a potent mitogenic factor that plays an important role in the growth, proliferation and differentiation of numerous cell types. This protein acts by binding the high affinity cell surface receptor, epidermal growth factor receptor. Defects in this gene are the cause of hypomagnesemia type 4. Dysregulation of this gene has been associated with the growth and progression of certain cancers. Alternate splicing results in multiple transcript

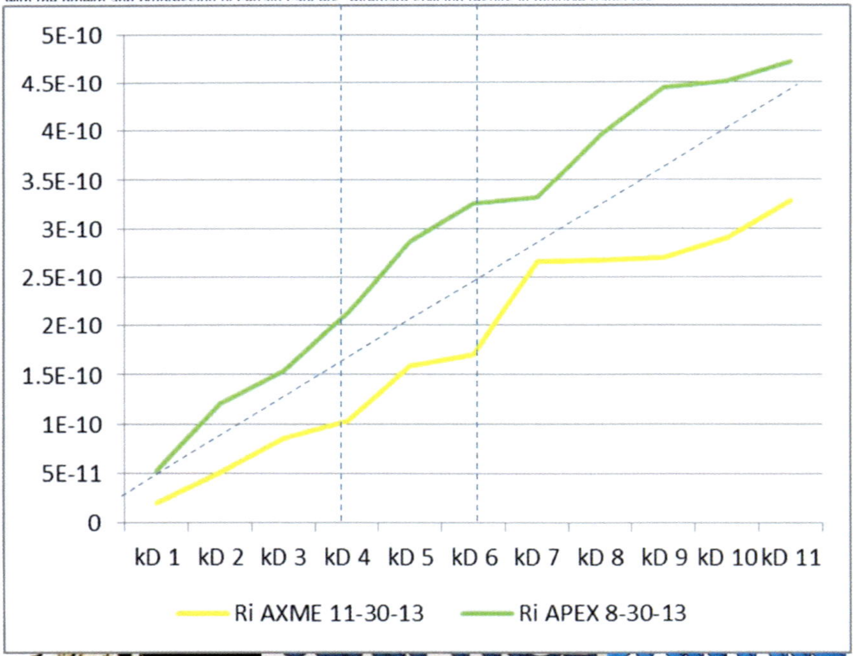

"and for that matter,
to what extent that which lies ahead
is already behind us."

US007598049B2

$y = 0.25\,x + .000125$

(12) **United States Patent**
Ray et al.

(10) Patent No.: US 7,598,049 B2
(45) Date of Patent: Oct. 6, 2009

National Institute of Genetics
Research Organization of Information and Systems

z Gene summary for CCL5 Gene:

This gene is one of several chemokine genes clustered on the q-arm of chromosome 17. Chemokines form a superfamily of secreted proteins involved in immunoregulatory and inflammatory processes. The superfamily is divided into four subfamilies based on the arrangement of the N-terminal cysteine residues of the mature peptide. This chemokine, a member of the CC subfamily, functions as a chemoattractant for blood monocytes, memory T helper cells and eosinophils. It causes the release of histamine from basophils and activates eosinophils. This cytokine is one of the major HIV-suppressive factors produced by CD8+ cells. It functions as one of the natural ligands for the ch
chemokine (C-C motif) receptor 5 (CCR5), and it suppresses in vitro replication of the R5 strains
CCR5 as a coreceptor. Alternative splicing results in multiple transcript variants that encode diffe
(provided by RefSeq, Jul 2013)

343.75

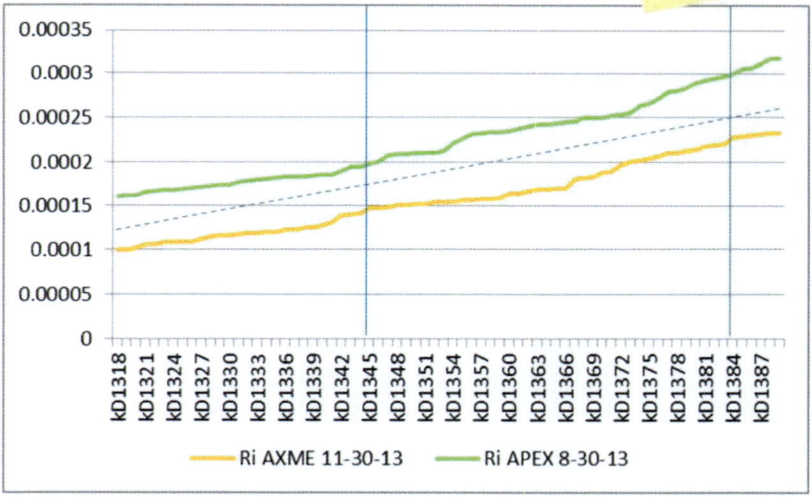

Les grands principes de la remise à plat des ouvertures de commerce le dimanche

מכון ויצמן למדע
WEIZMANN INSTITUTE OF SCIENCE

US007598049B2

National Autonomous University of Mexico
Universidad Nacional Autónoma de México
Official seal

(12) **United States Patent**
Ray et al.

(10) **Patent No.:** US 7,598,049 B2
(45) **Date of Patent:** Oct. 6, 2009

Platelet-Derived Growth Factor-BB (PDGF-BB) is a disulfide-linked homodimer of two B-chains (109 aa each). It is stored in platelet a-granules and is released folllowing activation of platelets. PDGF-BB has potent mitogenic effects on various cell types such as cells of connective tissue, bone and cartilage and smooth muscle. It promotes bone formation and soft tissue repair, stimulating wound healing processes. PDGF-BB is involved in the development of the nervous system and may act as a neuronal regulatory agent via nerve regeneration and proliferation. It also plays a role in the epithelial cell development of the respiratory system, cell migration and angiogenesis. **and Alzheimer's.**

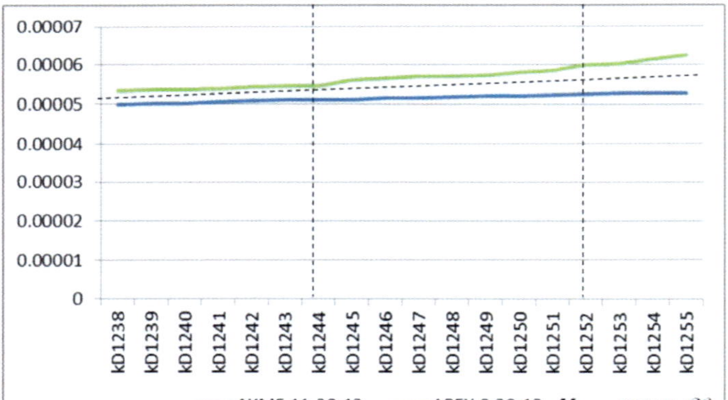

$$y = .03226\,x + .000052$$

3.23

FTSE
6,532.43
-62.90 (0.95%)

DAX
9,223.40
-178.56 (1.90%)

CAC 40
4,172.44
-113.37 (2.65%)

Virgin BOOKS

y = .453125 x + .0009

Read your stars every day with Mystic Me

CAMBRIDGE IMMUNOLOGY

US007598049B2

(12) **United States Patent**
Ray et al.

(10) **Patent No.:** **US 7,598,049 B2**
(45) **Date of Patent:** **Oct. 6, 2009**

38.61

Gene summary for CNTF Gene:
The protein encoded by this gene is a polypeptide hormone whose actions appear to be restricted to t where it promotes neurotransmitter synthesis and neurite outgrowth in certain neuronal populations. Th potent survival factor for neurons and oligodendrocytes and may be relevant in reducing tissue destructi inflammatory attacks. A mutation in this gene, which results in aberrant splicing, leads to ciliary neurotr

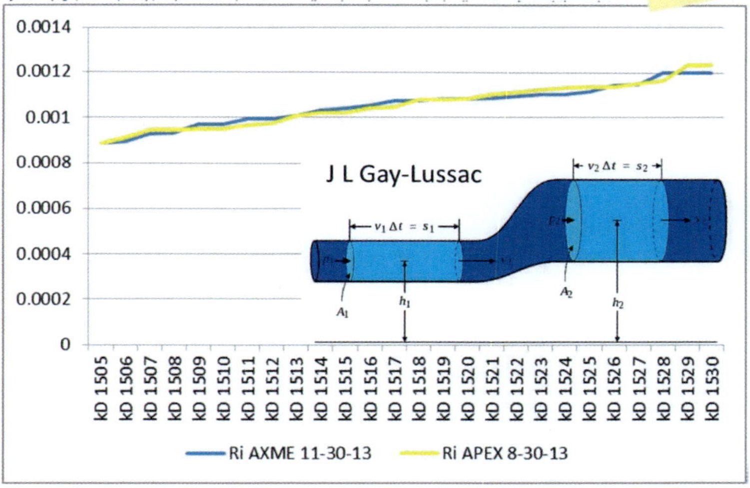

TOWIE's Chloe makes a showie of herself after night out
Lucy drops the bombshell and does a bowie out of Towie

Universitat Autònoma de Barcelona

UNIVERSITAS INDONESIA

Veritas, Probitas, Iustitia Est 1849

US007598049B2

(12) **United States Patent** (10) **Patent No.: US 7,598,049 B2**

Ray et al. (45) **Date of Patent: Oct. 6, 2009**

Gene summary for BDNF Gene:

The protein encoded by this gene is a member of the nerve growth factor family. It is induced by c... ...nd is necessary for survival of striatal neurons in the brain. Expression of this gene is reduced in both ...mer's and Huntington disease patients. This gene may play a role in the regulation of stress response and in the biology of mood disorders. Multiple transcript variants encoding distinct isoforms have been described for this gene. (provided by RefSeq, Jan 2009)

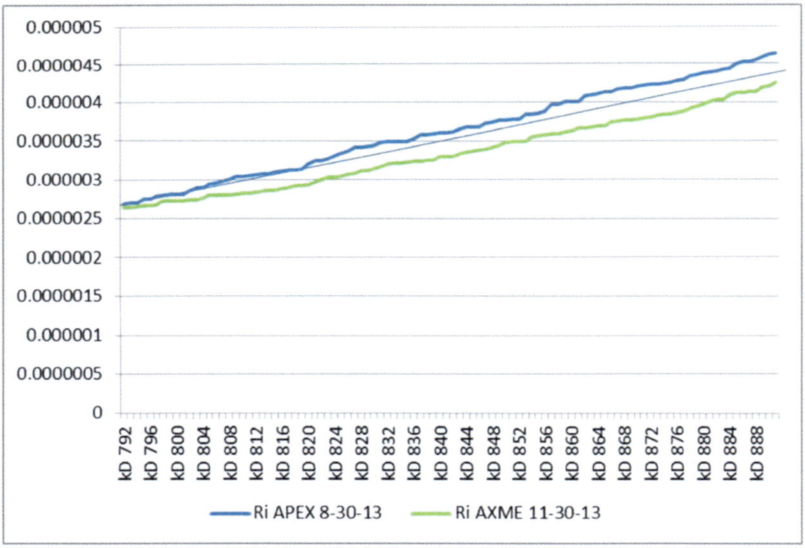

y = **.03226** x + **.000052** la vida es sobre todo un deporte de espectadores, de todos modos . . .

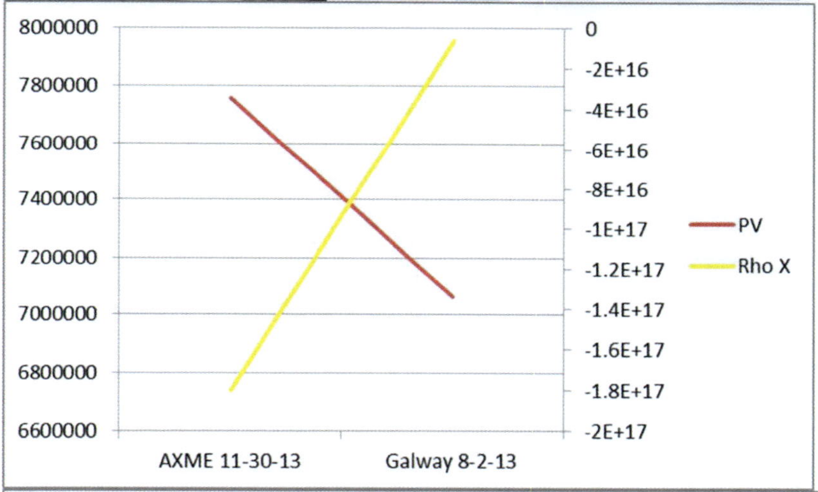

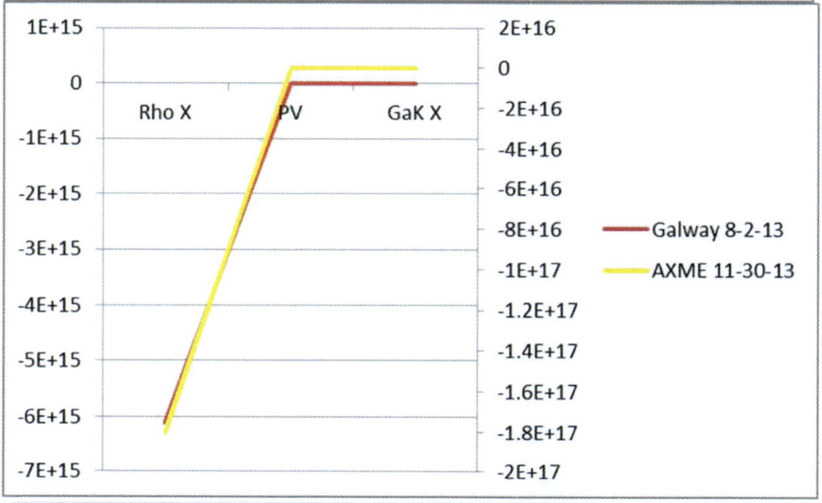

Gold 1,230.20 -1.36% Brent Crude 110.40 -0.65% GBP/USD 1.6337 -0.28% EUR/GBP 0.8370 +0.90%

Carter
A style original
Timeless fashion fixes

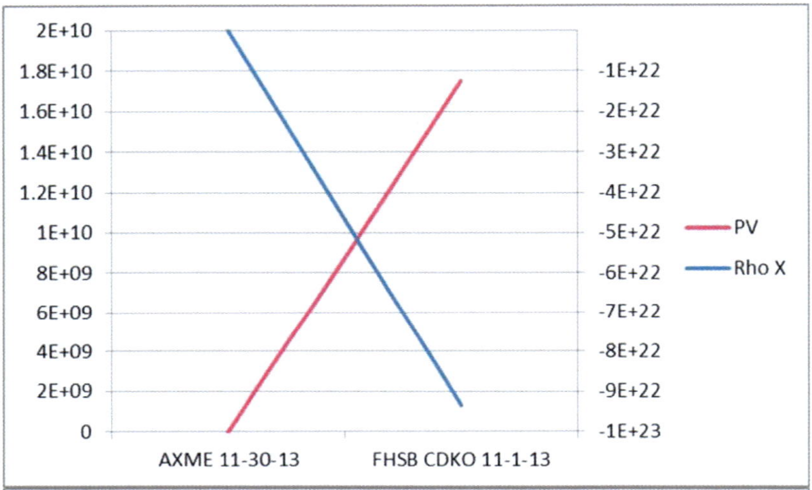
2E+10
1.8E+10
1.6E+10
1.4E+10
1.2E+10
1E+10
8E+09
6E+09
4E+09
2E+09
0
-1E+22
-2E+22
-3E+22
-4E+22
-5E+22
-6E+22
-7E+22
-8E+22
-9E+22
-1E+23
AXME 11-30-13
FHSB CDKO 11-1-13
PV
Rho X

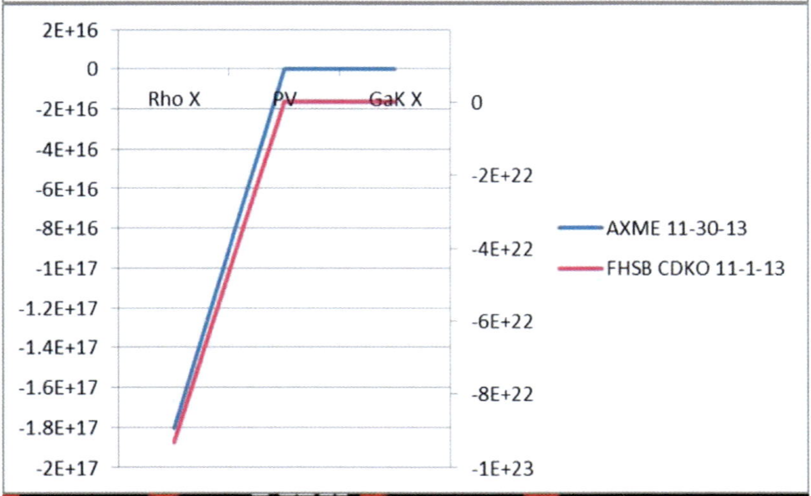
2E+16
0
-2E+16
-4E+16
-6E+16
-8E+16
-1E+17
-1.2E+17
-1.4E+17
-1.6E+17
-1.8E+17
-2E+17
Rho X
PV
GaK X
0
-2E+22
-4E+22
-6E+22
-8E+22
-1E+23
AXME 11-30-13
FHSB CDKO 11-1-13

THE DAY OF THE JACKAL

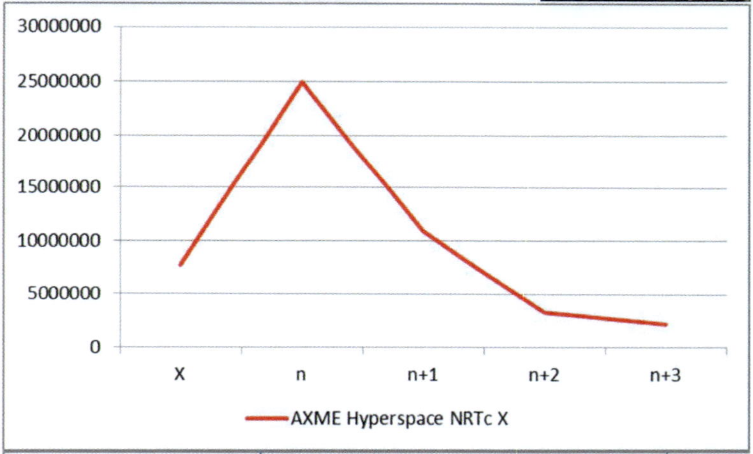

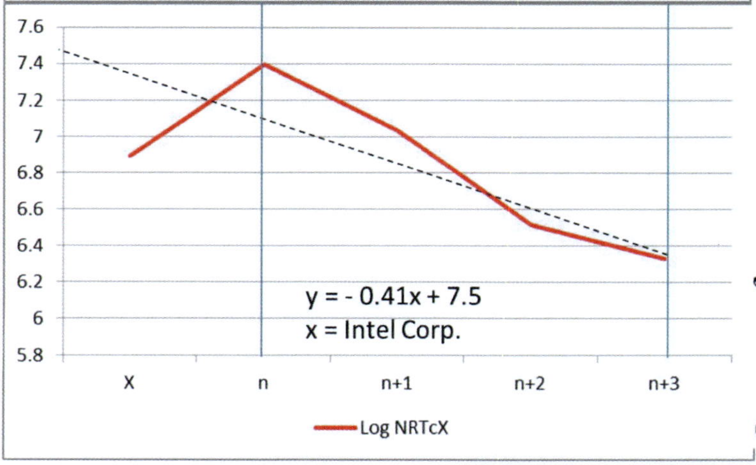

New York train derailment kills four, injures 63

Que faire à Paris ce week-end ?

TEXAS INSTRUMENTS

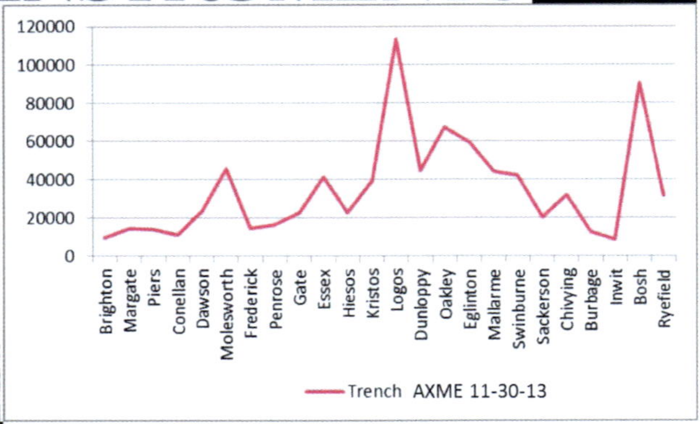

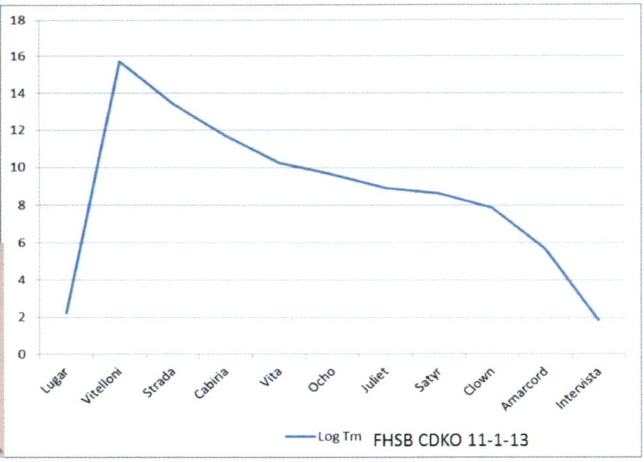

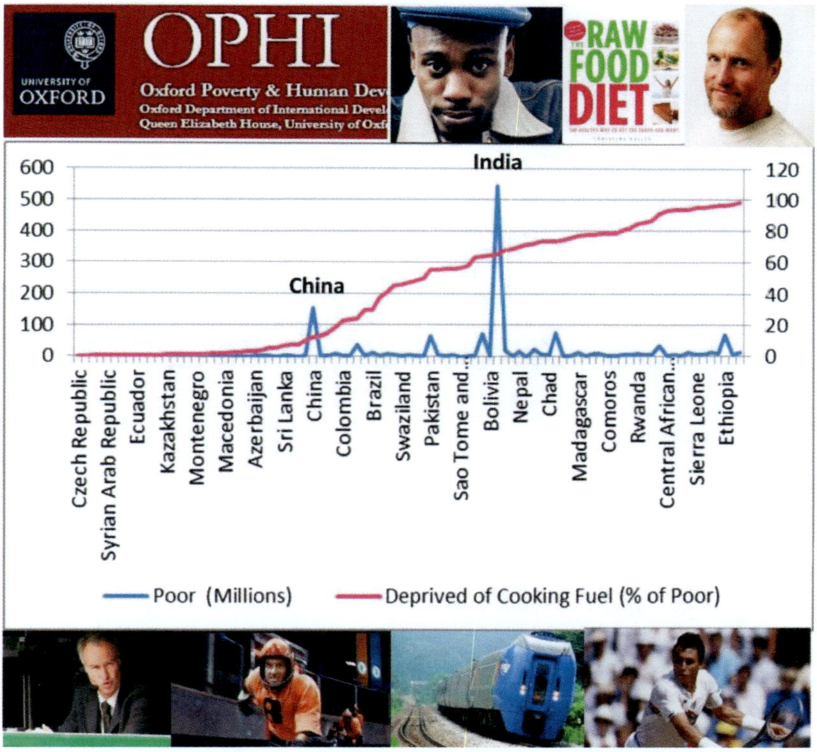
UNIVERSITY OF OXFORD
OPHI
Oxford Poverty & Human Dev
Oxford Department of International Devel
Queen Elizabeth House, University of Oxf
RAW FOOD DIET
India
China
600
500
400
300
200
100
0
120
100
80
60
40
20
0
Czech Republic
Syrian Arab Republic
Ecuador
Kazakhstan
Montenegro
Macedonia
Azerbaijan
Sri Lanka
China
Colombia
Brazil
Swaziland
Pakistan
Sao Tome and...
Bolivia
Nepal
Chad
Madagascar
Comoros
Rwanda
Central African...
Sierra Leone
Ethiopia
Poor (Millions)
Deprived of Cooking Fuel (% of Poor)

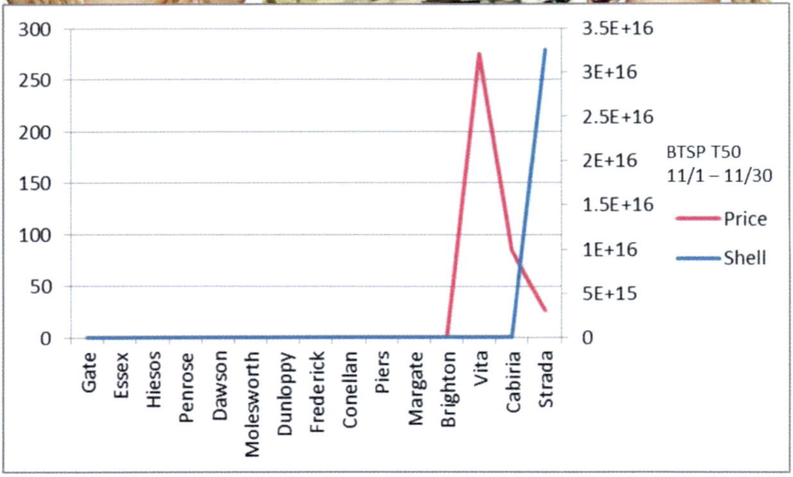

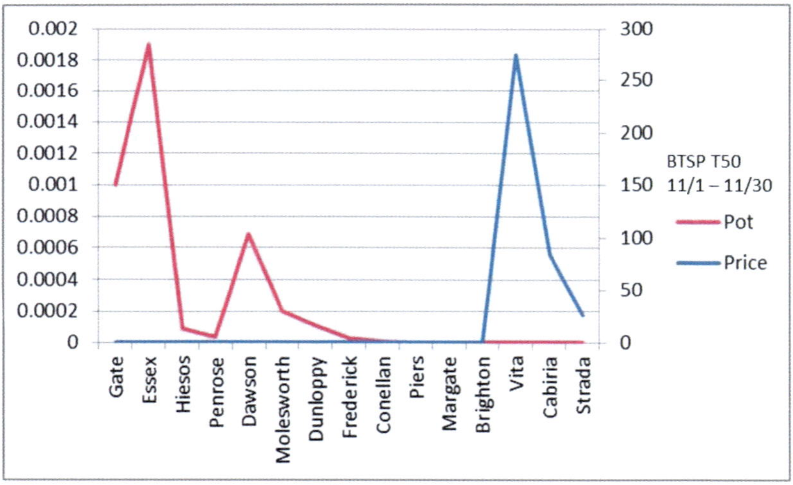

L'OM se sépare d'Élie Baup

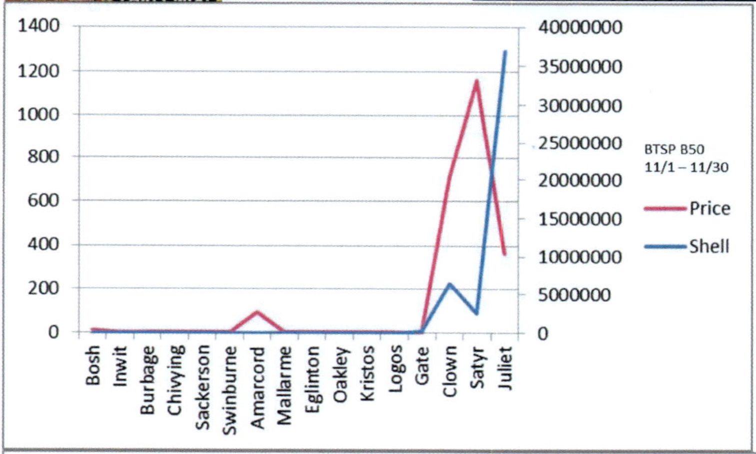

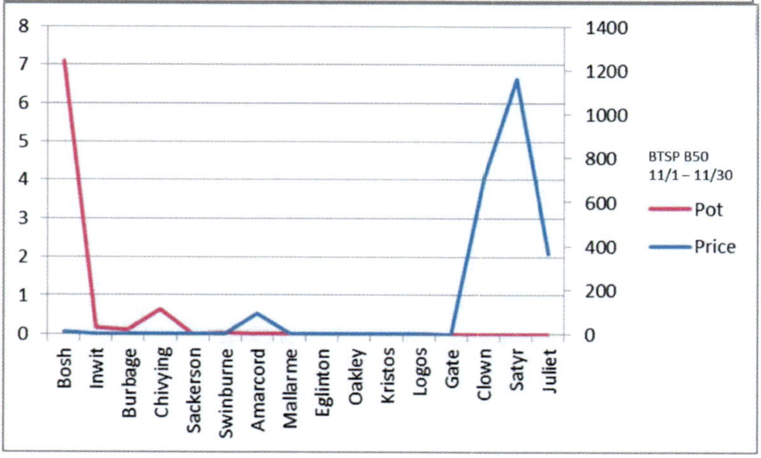

Rire de soi pour s'en sortir

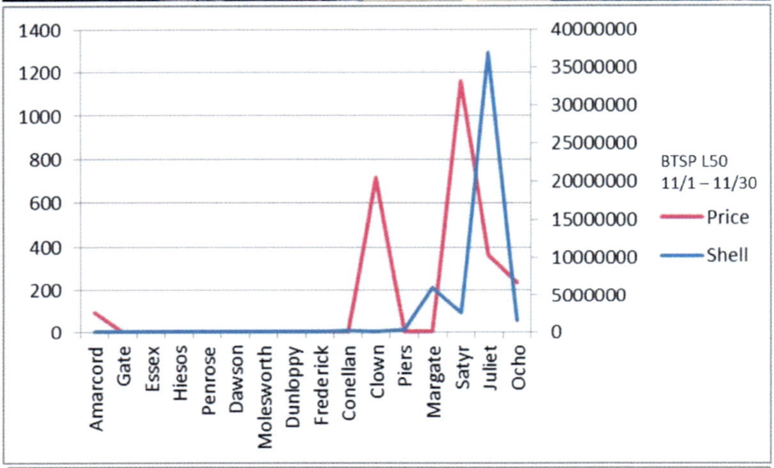

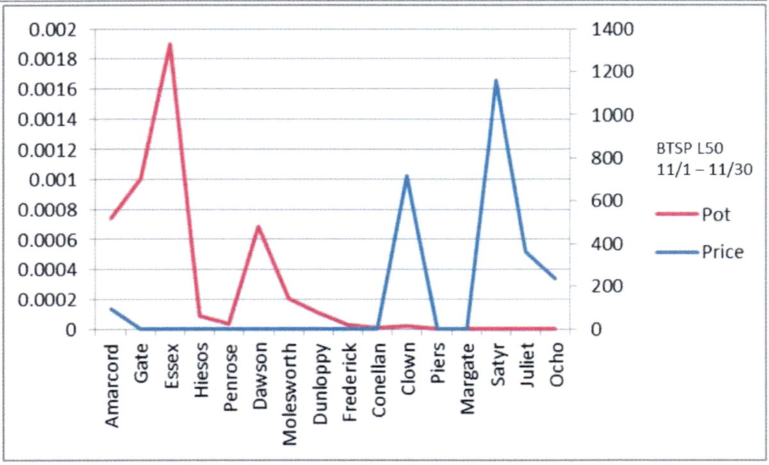

La Maison de Epinal, Vosges

COMBIN

ME THEORY

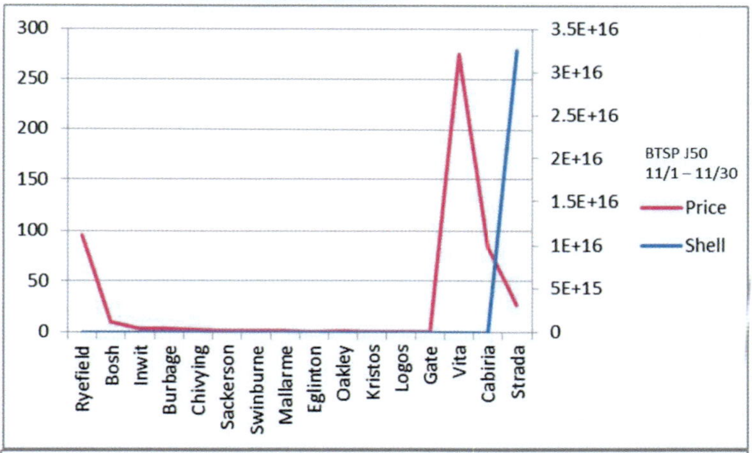

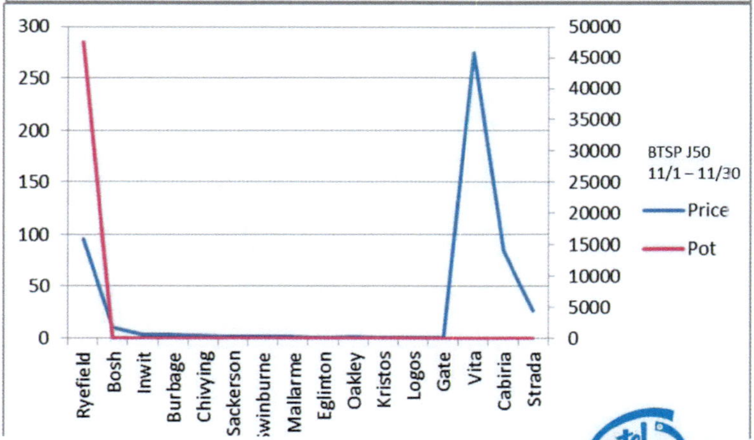

vos casseroles numériques

Gérard Depardieu annonce son retour au intel inside théâtre

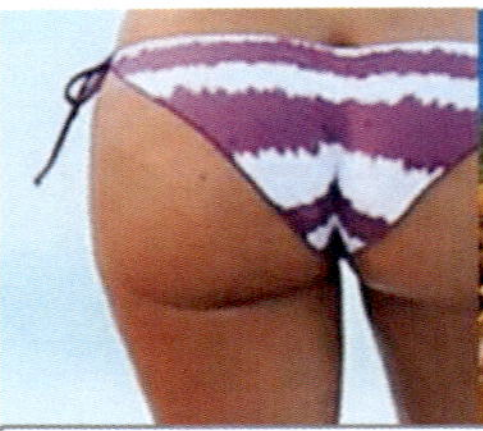

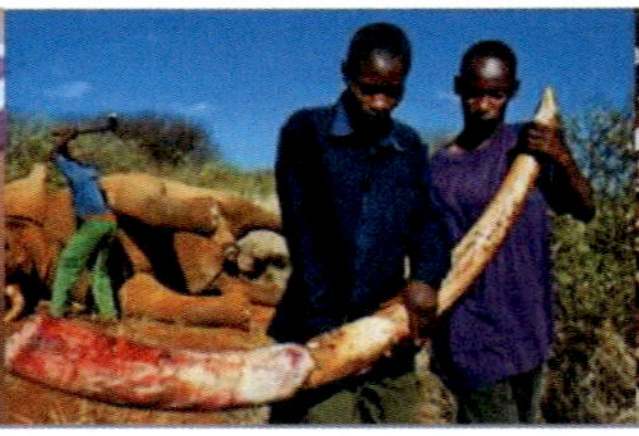

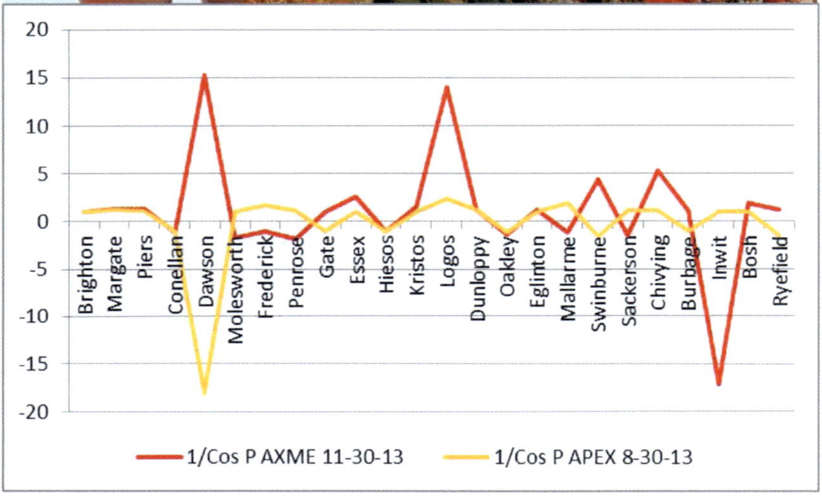
20
15
10
5
0
-5
-10
-15
-20
Brighton
Margate
Piers
Conellan
Dawson
Molesworth
Frederick
Penrose
Gate
Essex
Hiesos
Kristos
Logos
Dunloppy
Oakley
Eglinton
Mallarme
Swinburne
Sackerson
Chivying
Burbage
Inwit
Bosh
Ryefield
1/Cos P AXME 11-30-13
1/Cos P APEX 8-30-13

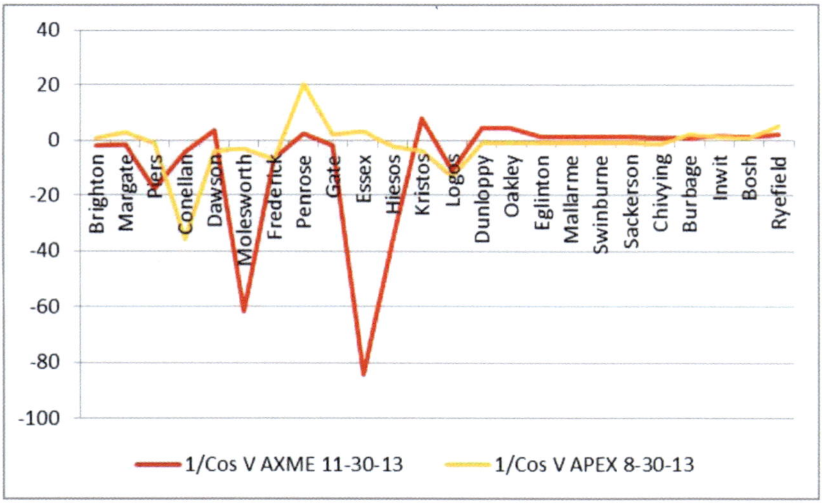
40
20
0
-20
-40
-60
-80
-100
Brighton
Margate
Piers
Conellan
Dawson
Molesworth
Frederick
Penrose
Gate
Essex
Hiesos
Kristos
Logos
Dunloppy
Oakley
Eglinton
Mallarme
Swinburne
Sackerson
Chivying
Burbage
Inwit
Bosh
Ryefield
1/Cos V AXME 11-30-13
1/Cos V APEX 8-30-13

James
Joyce
Dubliners

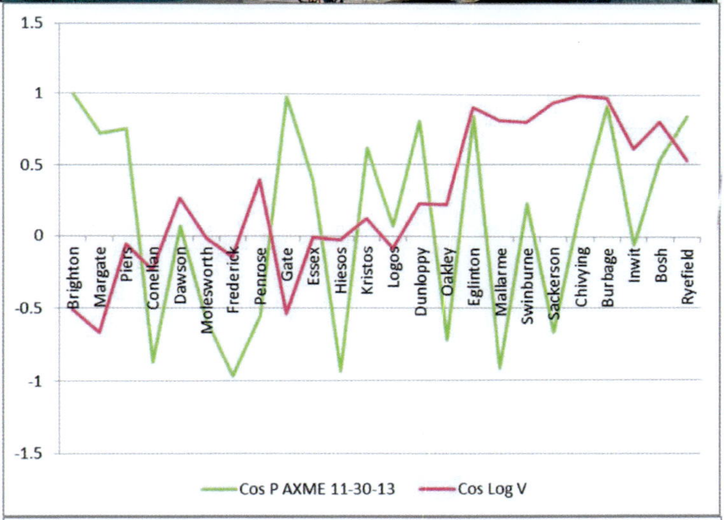
1.5
1
0.5
0
-0.5
-1
-1.5
Brighton
Margate
Piers
Conellan
Dawson
Molesworth
Frederick
Penrose
Gate
Essex
Hiesos
Kristos
Logos
Dunloppy
Oakley
Eglinton
Mallarme
Swinburne
Sackerson
Chivying
Burbage
Inwit
Bosh
Ryefield
Cos P AXME 11-30-13
Cos Log V

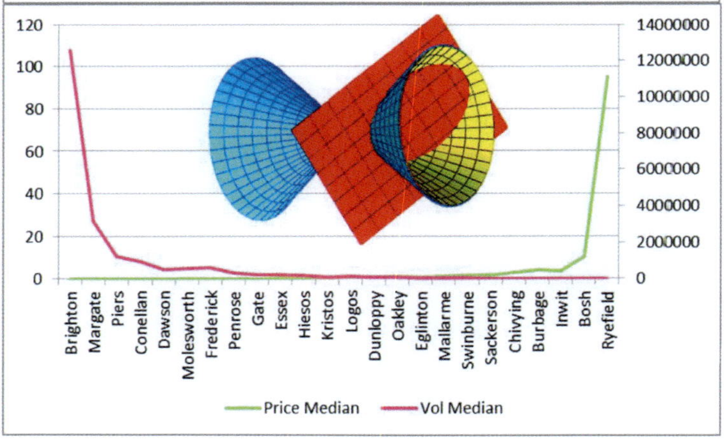
120
100
80
60
40
20
0
14000000
12000000
10000000
8000000
6000000
4000000
2000000
0
Brighton
Margate
Piers
Conellan
Dawson
Molesworth
Frederick
Penrose
Gate
Essex
Hiesos
Kristos
Logos
Dunloppy
Oakley
Eglinton
Mallarme
Swinburne
Sackerson
Chivying
Burbage
Inwit
Bosh
Ryefield
Price Median
Vol Median

MIT Technology Review

Detecting Alzheimer's Disease

A blood test that identifies the disorder in its early stages could soon be made available to researchers.

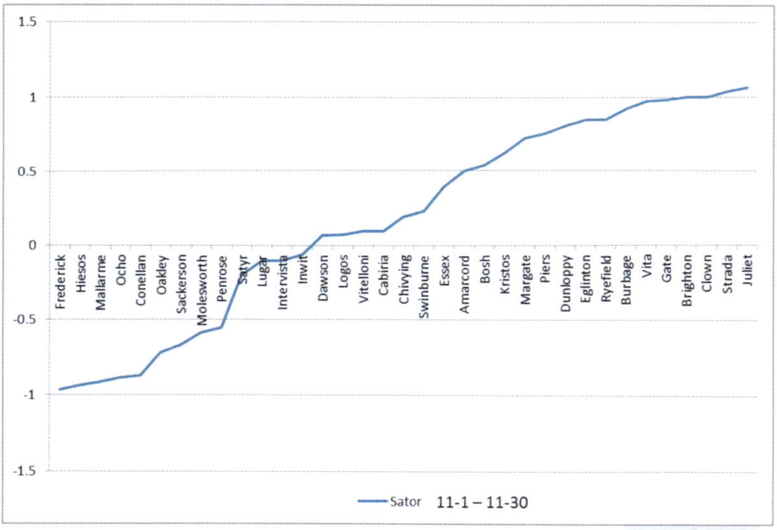

Satoris seeks $5M for Alzheimer's test — with high errors

VB

Business Wire
A Berkshire Hathaway Company

Satoris raises $5M, preps Alzheimer's test for research use

Satoris Receives Patent Allowance for Use of Plasma Biomarkers in Alzheimer's Disease Detection

Satoris Launches 3 New Blood Tests to Help Researchers Identify Alzheimer's and Develop Drug Therapies

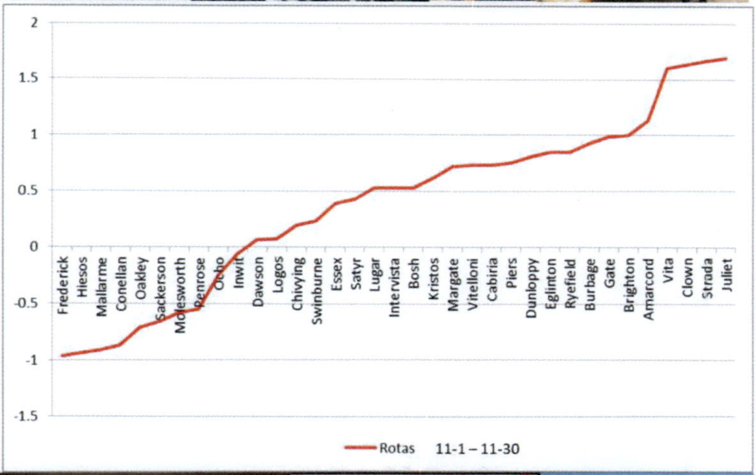

It may be that universal history is the history of a handful of metaphors.

Jorge Louis Borges

MERCHANT ADVENTURERS' HALL
YORK

La Grande Ligne est la séparation de la Seigneurie de Blainville, au Québec, en deux parties égales en 1792.

Thu, Dec 12, 2013, 7:01pm EST - US Markets Are Closed — Asia

NIKKEI 225	Hang Seng	SSE Comp
15,341.82	23,218.12	2,202.80
-173.24 (1.12%)	-120.12 (0.51%)	-1.37 (0.06%)

Gold 1,226.30 +0.11% Brent Crude 108.08 -1.14% USD/JPY 103.5750 -0.0048% USD/INR 61.85 +0.05%

FTSE	DAX	CAC 40
6,559.48	9,195.17	4,134.10
+7.49 (0.11%)	+22.76 (0.25%)	+4.73 (0.11%)

Gold 1,236.20 +0.59% Brent Crude 108.73 -1.20% GBP/USD 1.6423 +0.46% EUR/GBP 0.836

The Rolling Stones : 50 ans de rock en photo

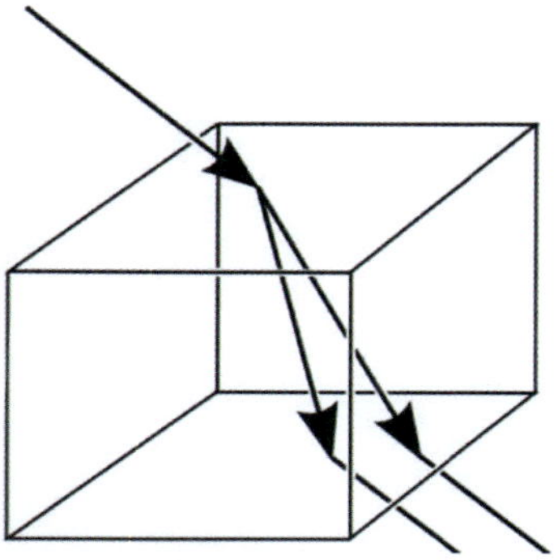

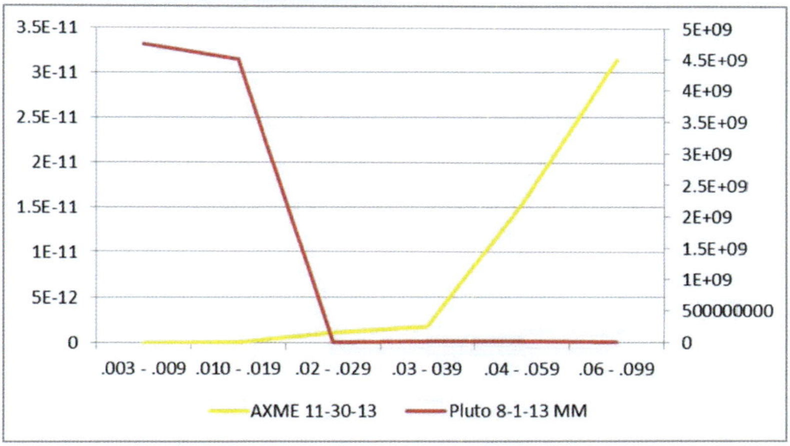

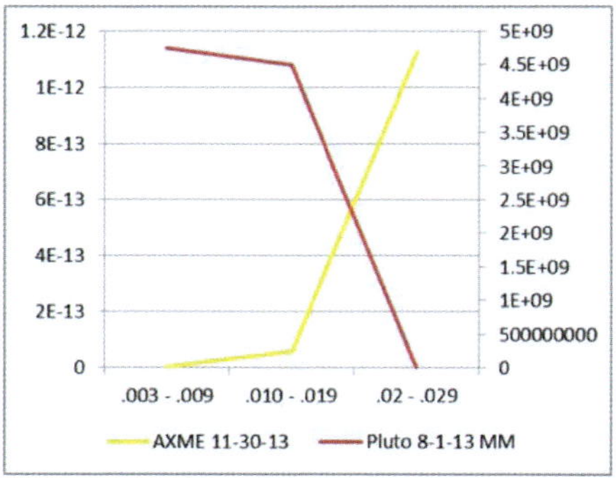

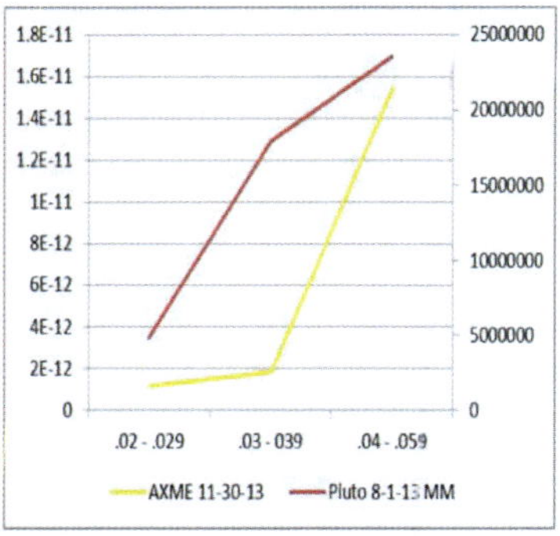

au débat sur la loi sur le voile intégral

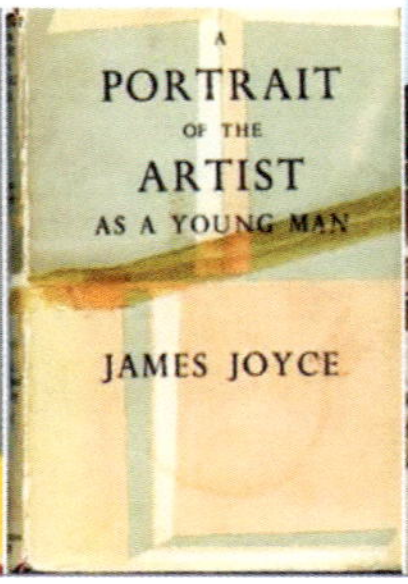

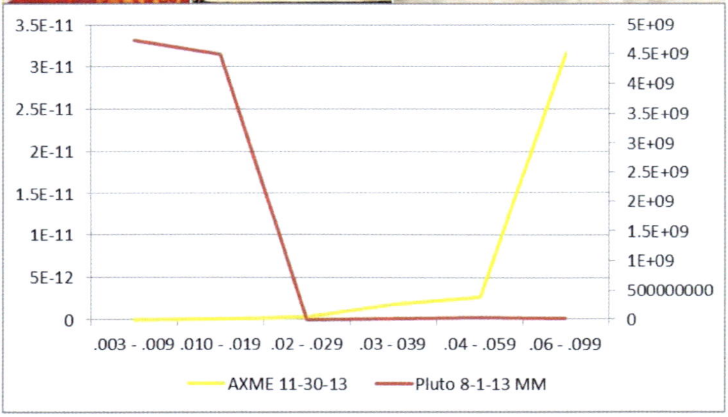

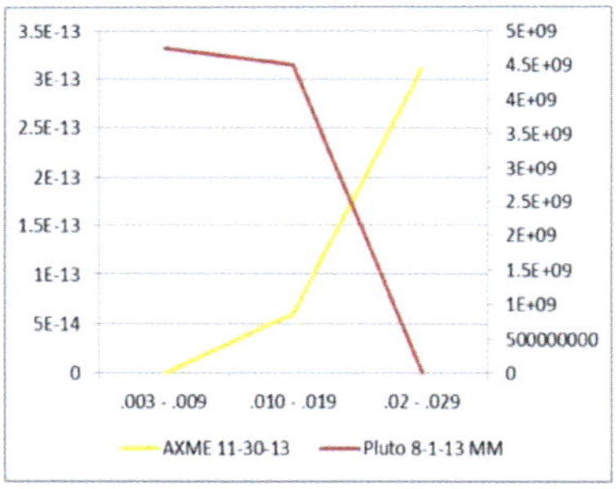

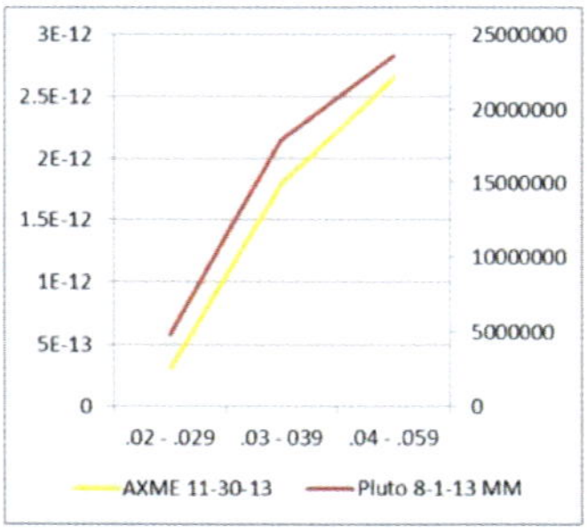

L'incroyable destin d'une référendum

GARCIA M

de una muerte

anunciada

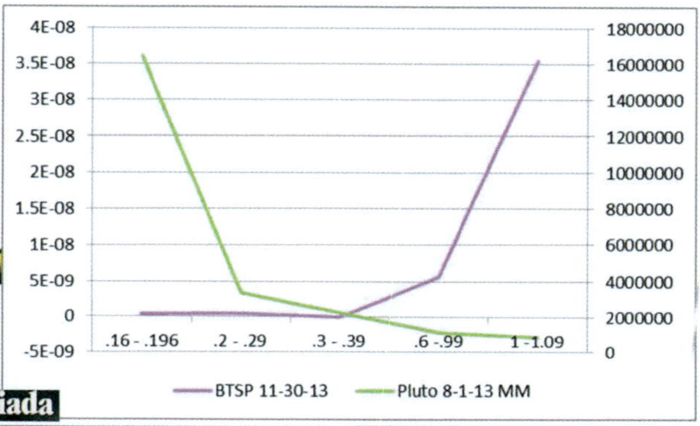

Crónica

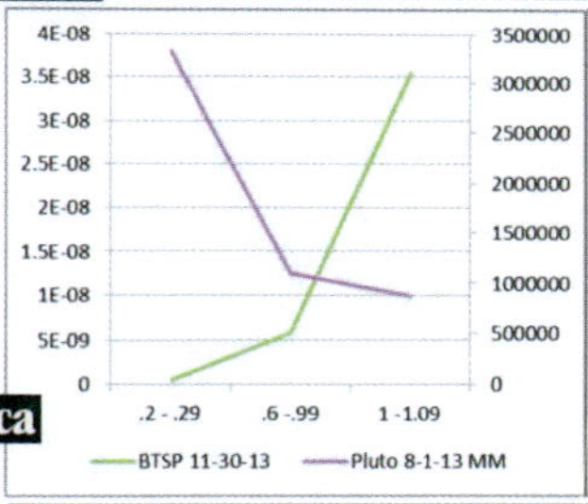

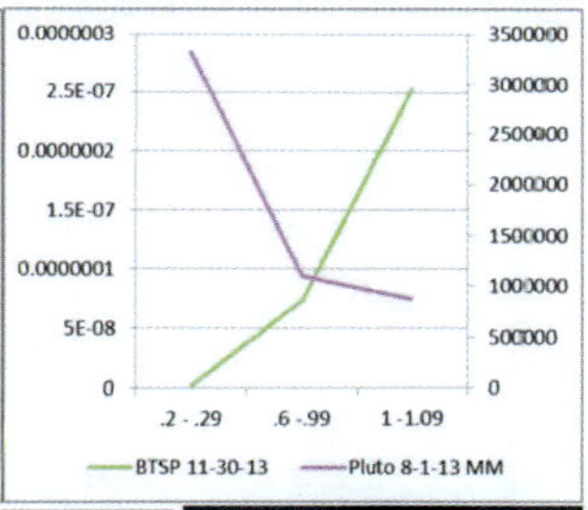

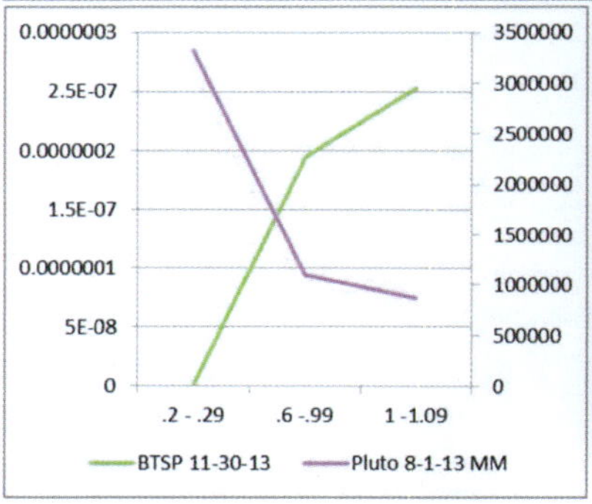

l'interprète en langue

Desjoyeaux, dans la Coupe de l'America

The Competitive Ochterlony Equation

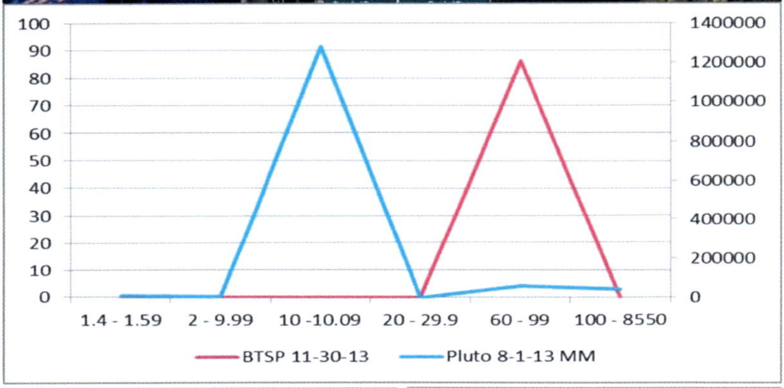

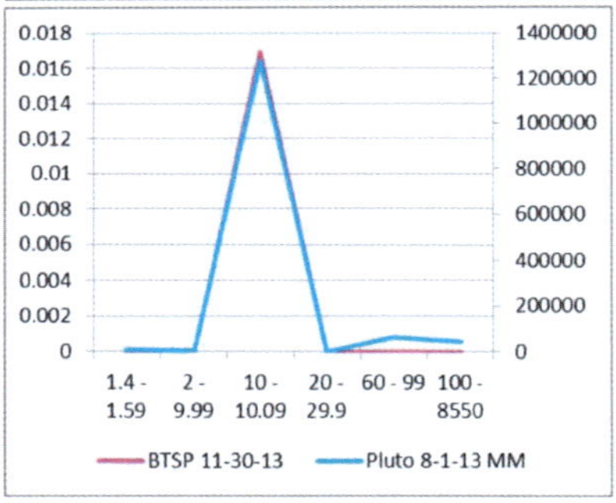

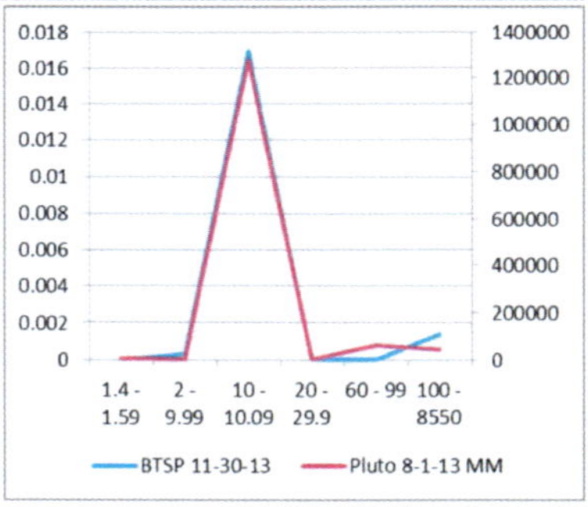

avec les partenaires sociaux

Dans les coulisses *Time*

Waldorf

Obviously, that is, though it may not appair to be sow tootha afferage policy wonks, the totality of the system is enclosed in layers which the ancient Chinese was pour potted leaf red to as the stratosphere and the ionosphere and the atmosphere of the Coppernican and Archemeridian institutions for spice trades along the Arctic passage for desypfering flux for the design of cascades in various markets as a means to azure proppa flu rates, rain fall being of the essence, except in the deseart where the manifold is hymnospheric or hollow for such indulgence as waterproofs and glosheshes or even so much assa dandylion in the medow as purview for sand and the brooding sea that doth washethith on too distant shores of its lapidary trails as fair inland as Dzur Dzuu from the Potomac dawn by the aroma of lapsang souchang tea in brewing Andes tea pots where one is likely to unearth the pacific heat of various alloys, steel say, versus aluminimum, for managing burn rates for poaching haggis on butter, holladaze, or street trash, for a swashbochlor's intangible needs as allaying fears of his managing partners in the event of market collapse, and it's been suggested by sources close to the executives at Rofl Dotch Shayl dat it mote bay a tad bid heapful four their lot to collect on debts owed by the MacKays if they were to be made aware off the landscape in front of them in broad brush stokes so as there is no confusion later as the swoshbocklors have delinked their infrastructure requirements and there is hardly likely to be an opportunity for the MacKays to corroborate around such circumstances as when the MacKays need guardrails gone opordone, you onnerterstan me Voutsinas, a communique of somesortorannor should be narrated to the MacKay's accountants, Bangor, I believe they are cold, who are the propietors of this hotel, with them snoopy chambermaids pilfering notes and snorkeling through disc drives like a boatload of focking lesbos who've adden't add their conts fucked by a long white one in decades, and their ass fucked desk clerks who have trouble with English enough times not to onnerstant da mey room's got no heat in the middle of facking winter at threehundred and fifty dollars a knot, besides the nammer of thames they keep calling my room from some focking nursing home in Sutton Place enquiring bot maturity dates for Dean Witter bonds, as judging by their obnoxious behaviors I would imagine that the Bangors fucked the MacKay's with a business plan which had no exit strategy, much like a coterie of inept bank robbers who managed to lock themselves inside a bank vault and are living in denial of getting out safe, acts of attrition, defalcate, rank, miscarriage and otter offences knocking at their doorstep with wagers on fast blood, what second, and when at lost, for pourparler. Across the street sitht the policy wonks for dare emergence, Voutsinas, troughs, canals, hitches, trolleys ready in the wake of shuttered manufacturing bases which could not met output costs for a future course which portends four miles downhill of thriple black diamonds with scaresely much room for hand holding or tangos lessons as the narrows are flush against the slope's ridge of steep inclines that chance one to ski likely into the ravine of dissipation than crashing into Mood's mountain side, and while accounting for a pulse field of moghuls along the way I should discourage the MacKays from carrying liquor bottles of rafid heat loss along the course inclusive for this holiday pachage, bleding what would result in thaumaturgic shock of halfing barely enough to float past the timber line into the pines without a lift in sight, speaking from experience, whence one ought not to conshider ambling as venue option with respect to the kifnetic pufftential and collision rates with Olympiads or for that matter loss of moss, for woth hath rothly earned the appellation, shall we say, Geneva Spur in the Alpines.

Can if you want, mention to Bangor, to scuttle out of Wonton, since it's likely to be swept under the rug like a dust ball, as their rental car agency does not amount to mooranan error for the scheme in place though I'd think they'd probably relish a demonstration befitting their ilk, playing big fish in a lyle pod, providing the circumstance art apt to cantilever the agency into the cherry orchard when the car clunks for beach badminton in simple whoremonics for damped oscillations with reassists by Kilkenny as umpire for an interesting doubles matchup between Beckworth, Shepard, Aniis, and Evylin, as she tends to be abstract in the formal sense of logic baring all on the performance of her Carlo, some say enemy of the people while others who'd rather contend with his hissyfits at betrayals than be left out of the table playing some street hoodlum who hasn't a cent to loose, intractable as such foils or as that Iago from Capulets to the Montagues among merchants of Venice, before playing the shuffleboard with etchasketch toys from Great Neck as a foist for a Swedish tease, there being nothing worse than a bunch of butt ugly fag ass niggers focking a Islamic bitch without a weenie, when the load voltage applied to her crotch does nosink to produce a pinch for the diamond industry as one would expect from a properly functioning bridge or wouldn't the drowning rats know how that works from oour ill diario concerting turtle under Briton, practically speakhin, or should perhaps suggest that they had been used as my linin shits, insulation for the day of the jackal, as sweet, treble, and a bit melancholy.

Waste land, waste land, O waste land, Voutsinas, single capacitor as load is purely reactive, poor going and coming, coming and going out of the inductor as the inductor temporarily gains and loses energy, with poor delivered to the magnitude of resistance in the circuit for a kite flyer's tenebrae in crossing hairs, tying knots, rigging the sails for the time machine's ostrich eggs, pass over death and decay, corrosion and bile, filth and swine parading across the balcony as degenerate warthogs of caprice, on the wings of a butterfly with a porch-lamp. The working capital went that way a year ago buying up seats in the business class and cargo holds while the chipmonks were making giddy with asteroids falling from the Kuiper belt, elementary school slaughter houses, mall wobberies, while wallowing in pierpfires for insurance reimbursements on nonworking assets in the fallowed hulls of agriculture for funding daycare for corn flakes and pop tarts in sharpening pencils against spiraling cost of tuition and diminishing availability of net present state subsidies amounting to thirty five thousand per pupil for the toilet papery industry while hurling insults at avaricious jews, gluttonous niggers, and wimpy weasel wogs from their Nazi palace of fockarray with FDA approved stroke pills, Viagra, Cialis, Levitra, for the storks, as a competitive business plan in a global marketplace and, yes another Knob Creek, please, what about yo, Voutsinas, that would be all then, wouldn't mind doing her behind the bar, that one, like a cervical pap smear and a bit of mammogram between me incisors, eh, Satan's cult of French letterist equivocating the rape of that Locke, sucking on hubris' fat cock for a platter of macaroni and cheese powder and some drivel on her nipple, as aperitif, of course.

Se habla español, douchebag?

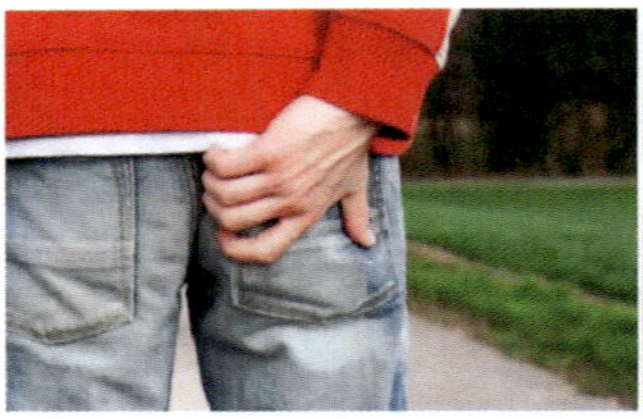

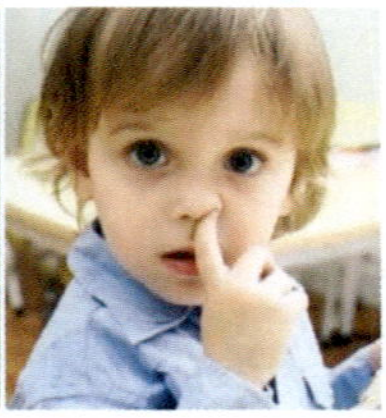

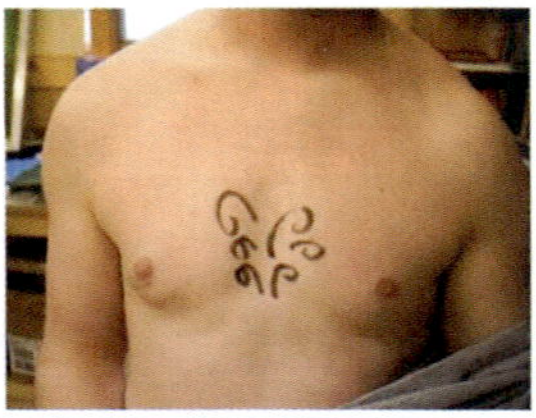

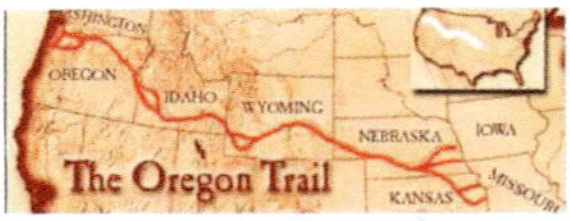

Parlez vous Francais, fuckface?

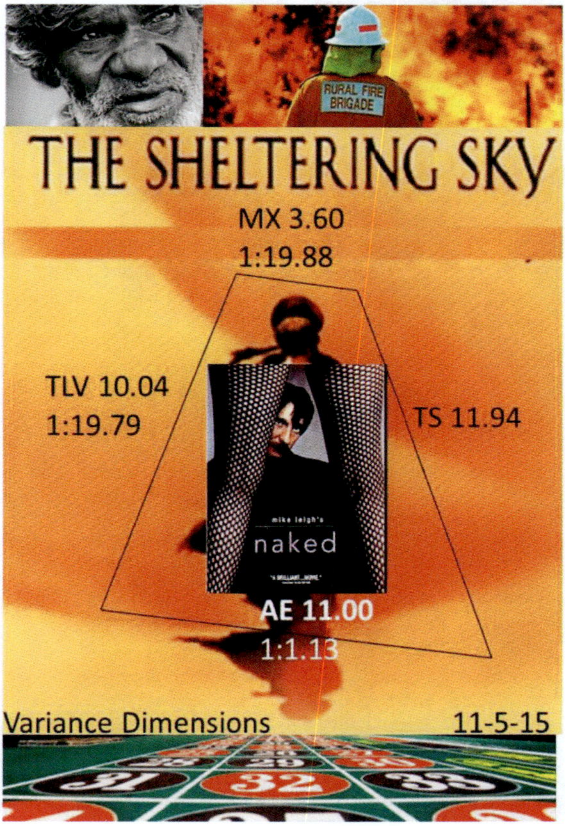

Mon. Dec 16. 2013. 4:00pm EST - US Markets Are Closed

Asia

NIKKEI 225	Hang Seng	SSE Comp
15,152.91	23,114.66	2,160.86
-250.20 (1.62%)	-131.30 (0.56%)	-35.21 (1.60%)

Gold 1,240.30 +0.46% Brent Crude 110.56 +1.59% USD/JPY 103.0075 -0.14% USD/INR 61.78 -0.67%

$$361.2\ \mathscr{Y} = 3659479;\ \mathscr{Y} = 10131.44$$

$$\mathscr{X} + \mathscr{Y} = 4.5 \times 10^{23}$$

Aislin
Cartoon of the day

HOME NEWS OPINION BUSINESS SPORTS ARTS LIFE HEALTH TECHNOLOGY

Stolen cobalt-60 finally removed from Mexico cornfield by robot a week after thieves dumped it

S&P 500	Dow	NASDAQ
1,786.54	15,884.57	4,029.52
+11.22 (0.63%)	+129.21 (0.82%)	+28.54 (0.71%)

FTSE	DAX	CAC 40
6,522.20	9,163.56	4,119.88
+82.24 (1.28%)	+157.10 (1.74%)	+60.17 (1.48%)

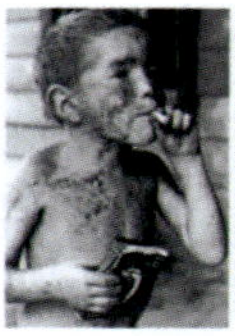 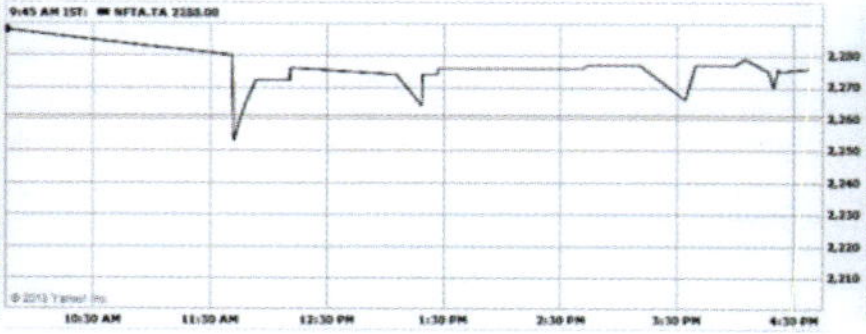 

1 — A 2.64 — 4312

100 — M 0.01 — 4312

800 — M 0.02 — 4312

1000 — M 0.03 — 4312

1250 — M 0.04 — 4312

1500 — M 0.05 — 4312

1750 — M 0.06 — 4312

1769 — M 0.10 — 4312

2400 — M 0.11 — 4312

2500 — M 0.15 — 4312

2800 — M 0.20 — 4312

3000 — M 0.30 — 4312

3500 — M 0.40 — 3741

3365 — M 0.50 — 4312

3740 — M 0.50 — 4100

3740 — M 0.60 — 4024

3841 — M 0.62 — 4024

3741 — M 1.0 — 4312

4024 — M 2.0 — 4312

4157 — M 3.0 — 4312

Regards Beaubourg

Welcome to Satanville

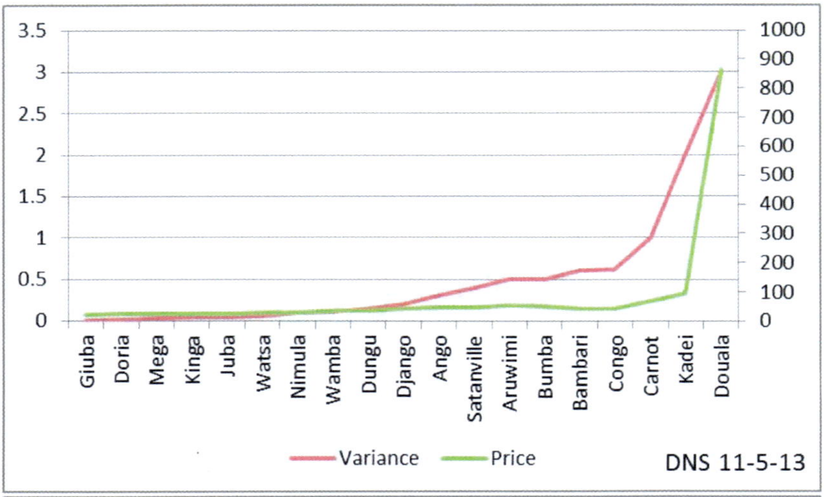

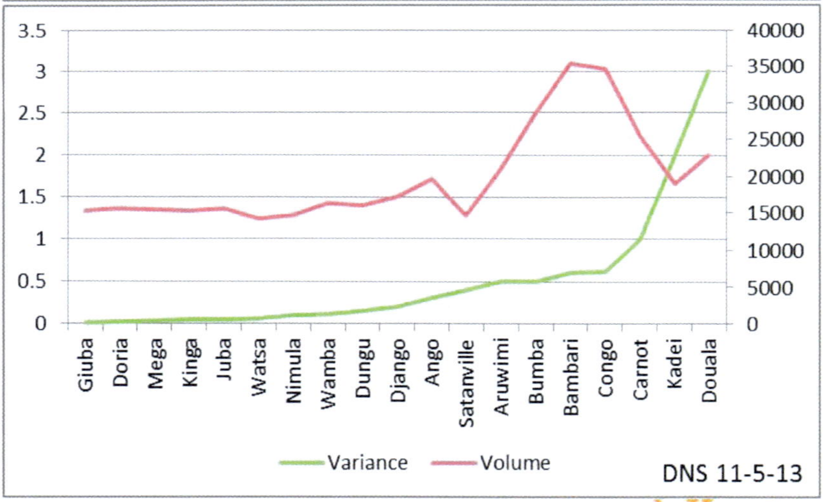

Les nouvelles représentations voyage à Kyoto

13andMe

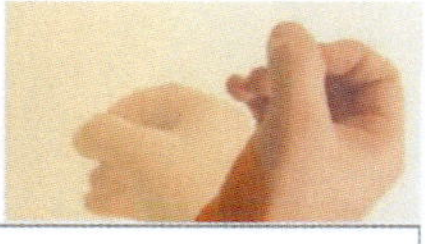

La Russie tend les bras à l'Ukraine location de smartphones

cadeaux droite côté

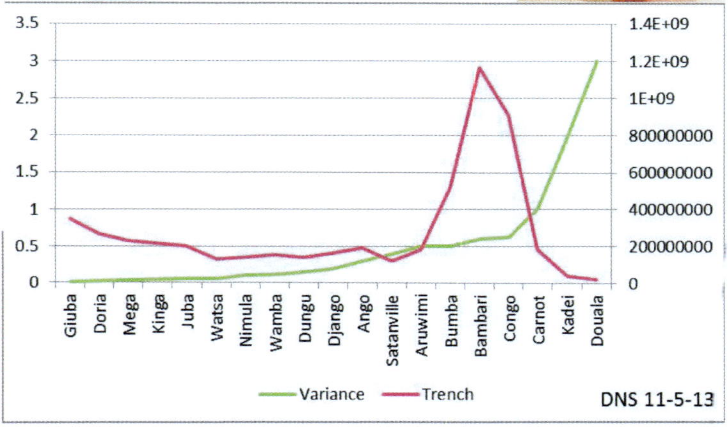

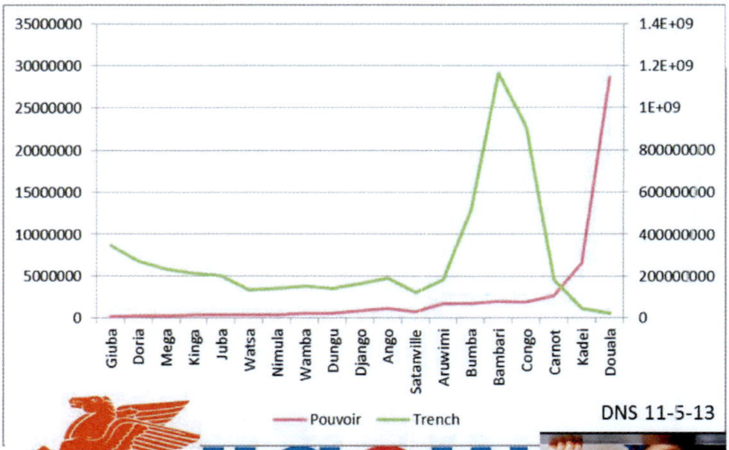

Mobil

À la découverte des Étrusques

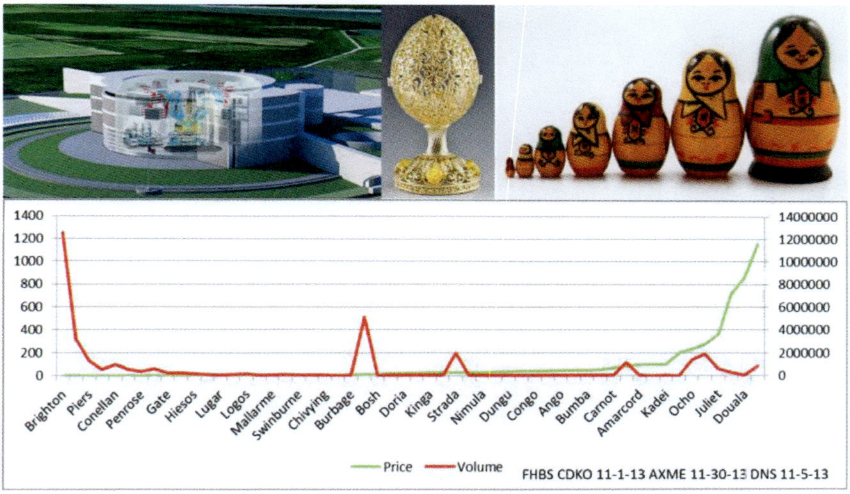

Comment le fisc nous surveille

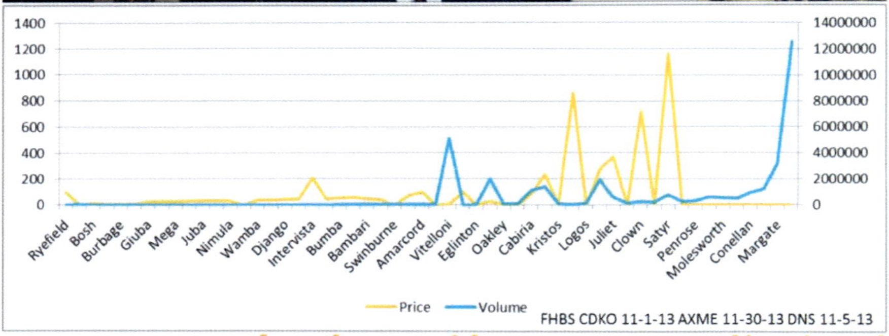

bouleversé le paysage audiovisuel

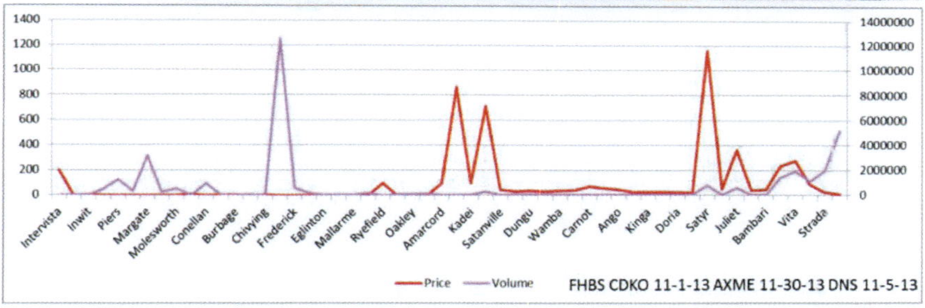

Révolution technologique dans les pacemakers

Fromages de Noël : les stars du plateau

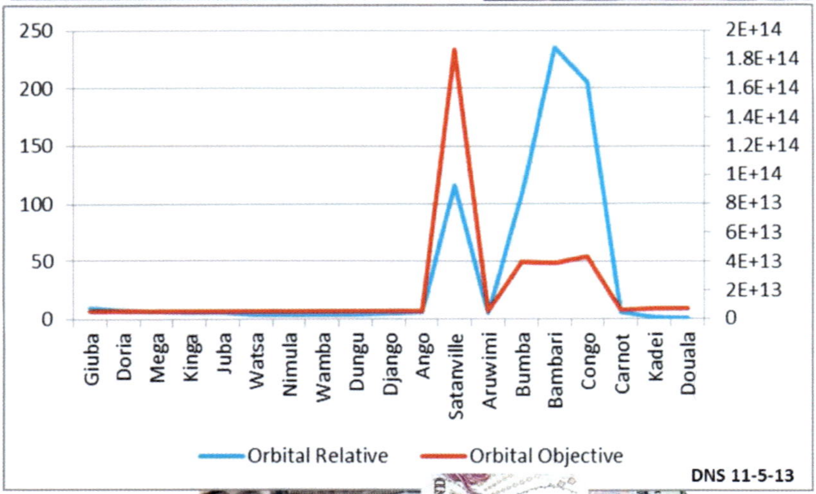

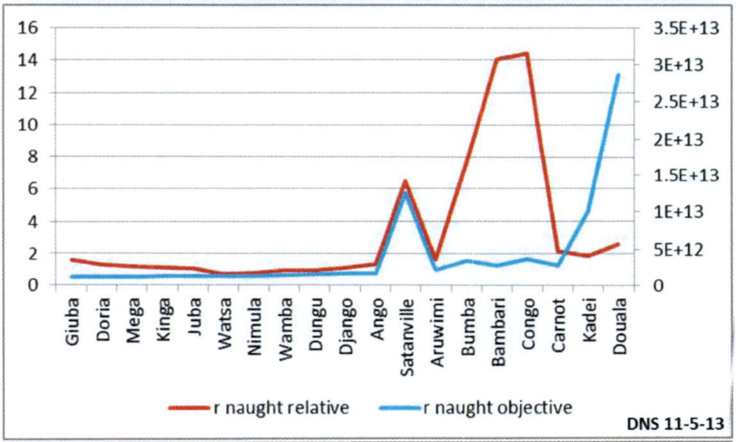

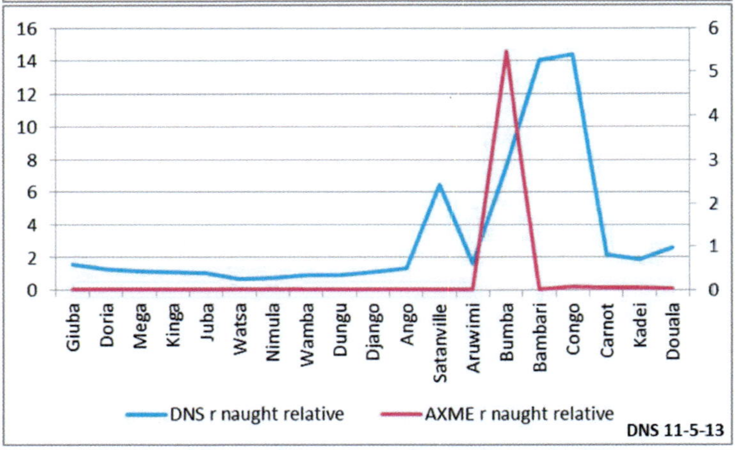

Lens, Metz, Marseille... lendemains de fête

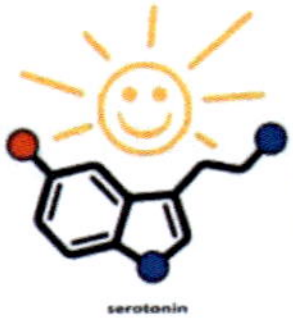

Dough fakir door, whare the doors,
Way a drop op smolten tonne
Medow a name call my chump
Fa along wong way to, whoa . . .

w. shapespeer

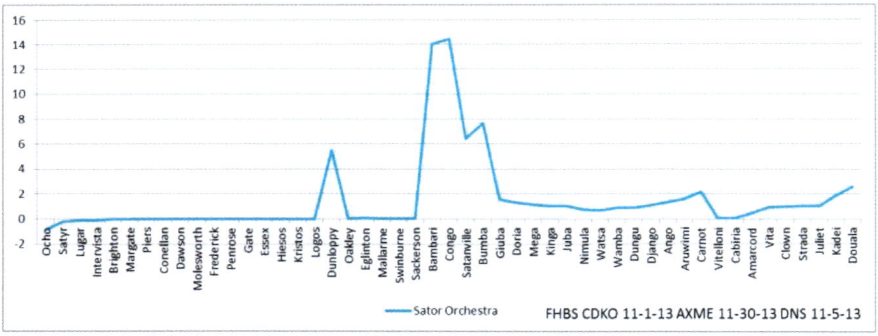

Joaquin Phoenix pourrait Lex Luthor

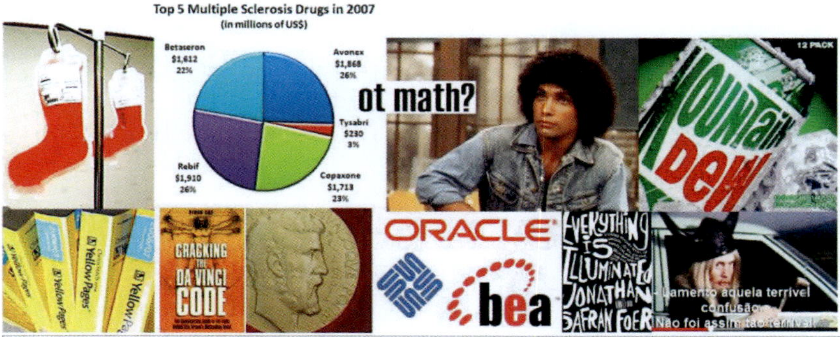

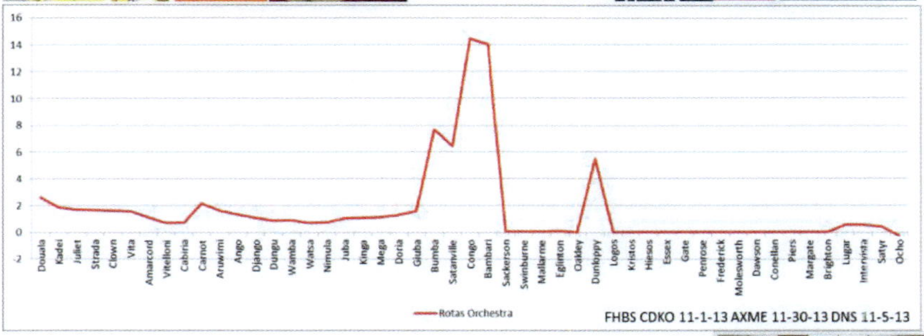

Deux trains rivaux pour desservir l'aéroport

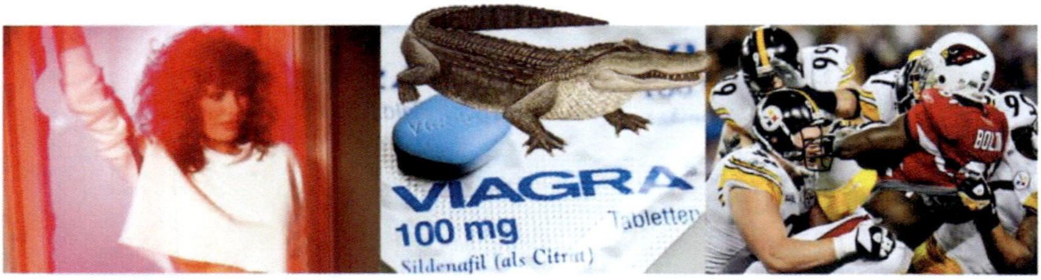

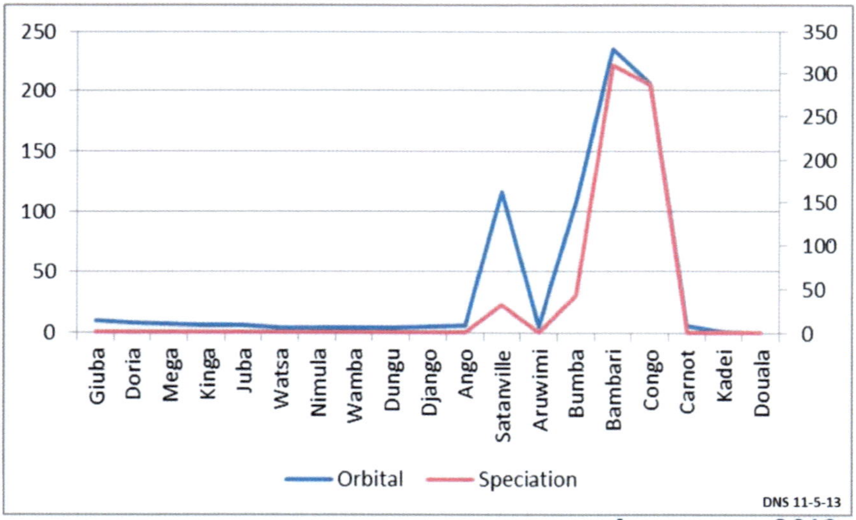

Fonctionnaires : les effectifs sont repartis à la hausse en 2012

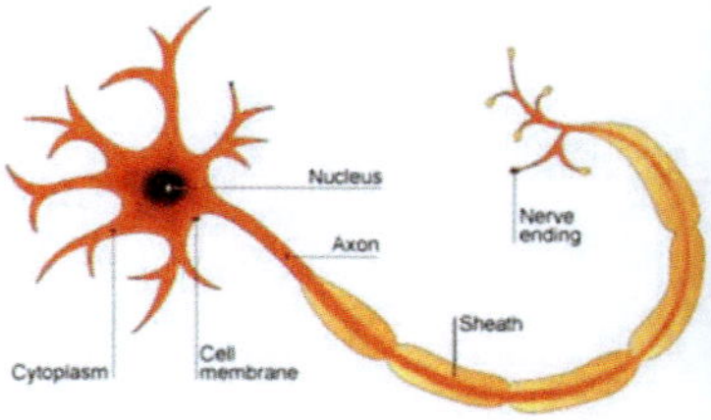

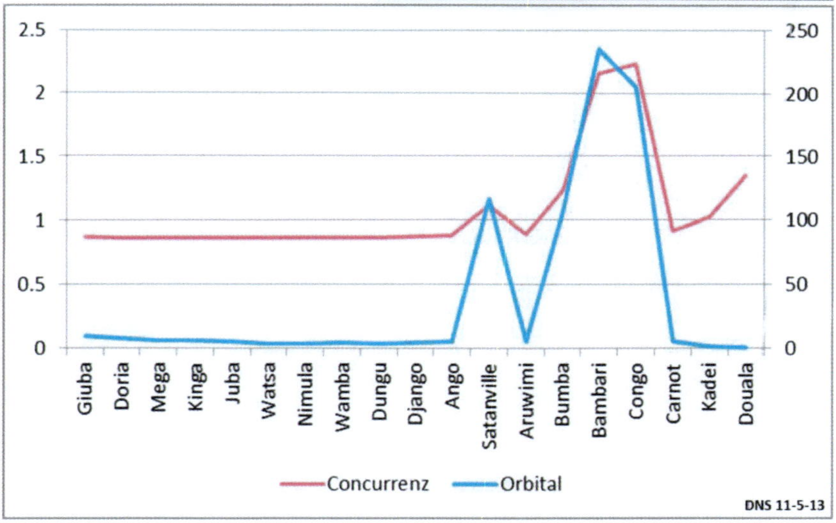

L'éolien, première énergie en Espagne

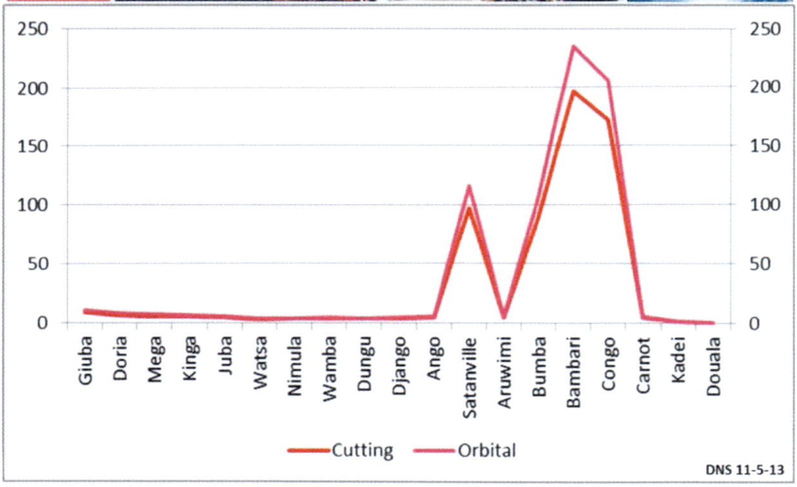

Les pires fashion faux pas de l'année

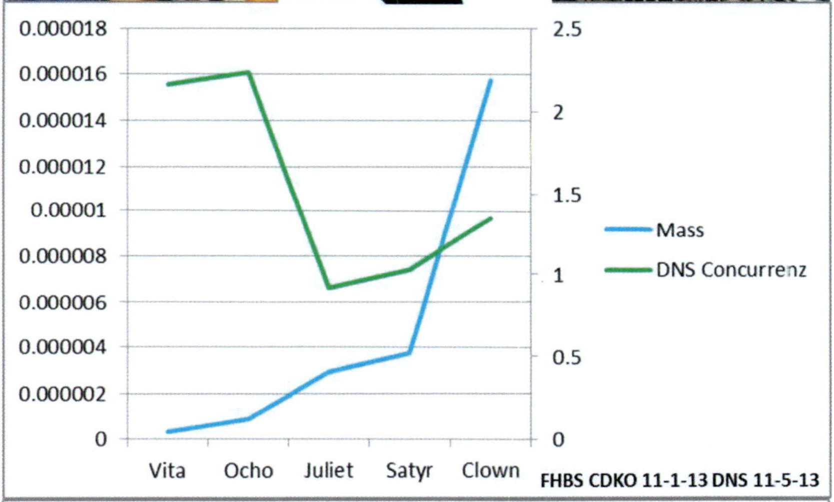

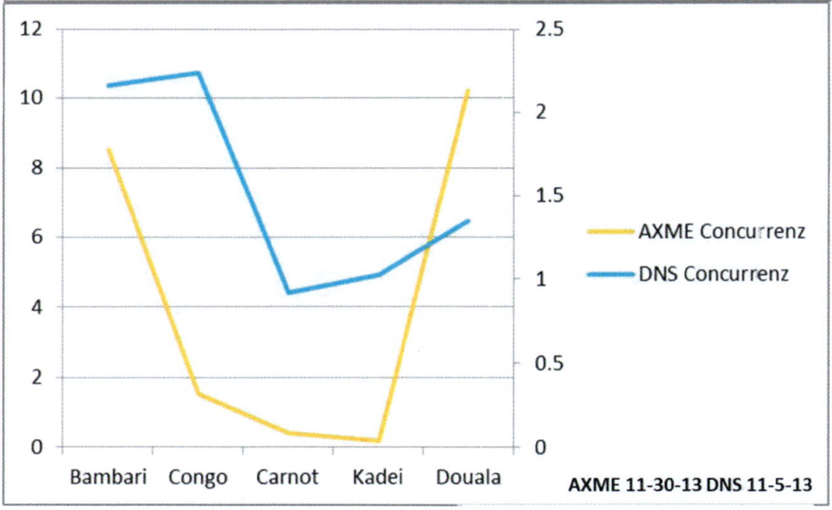

Hollande et l'Algérie, de la ouvriers du textile

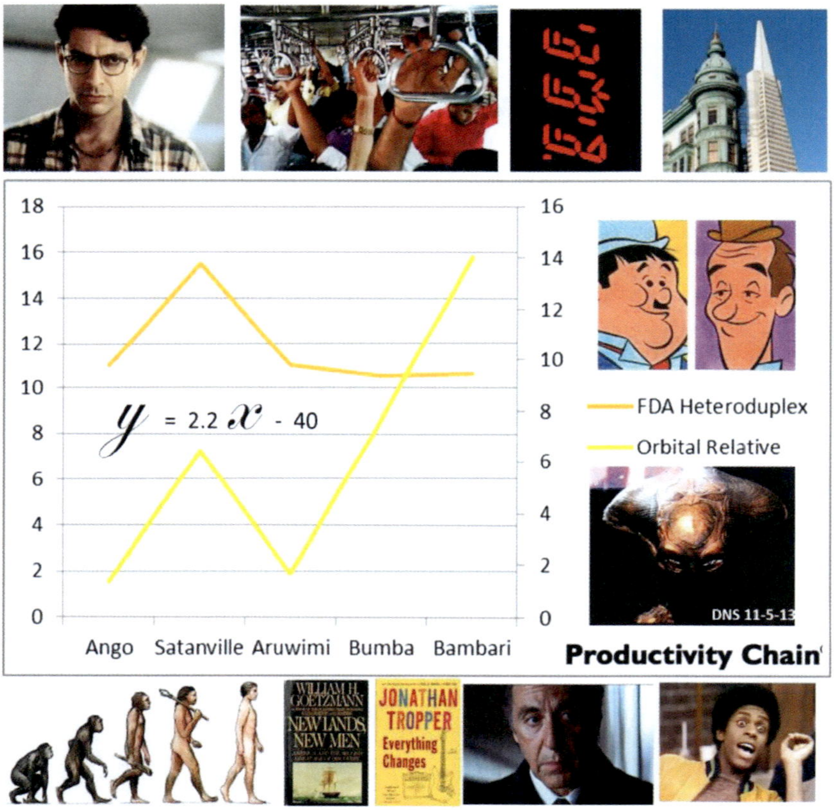

Khodorkovski veut mener un nouveau combat

a space odyssey

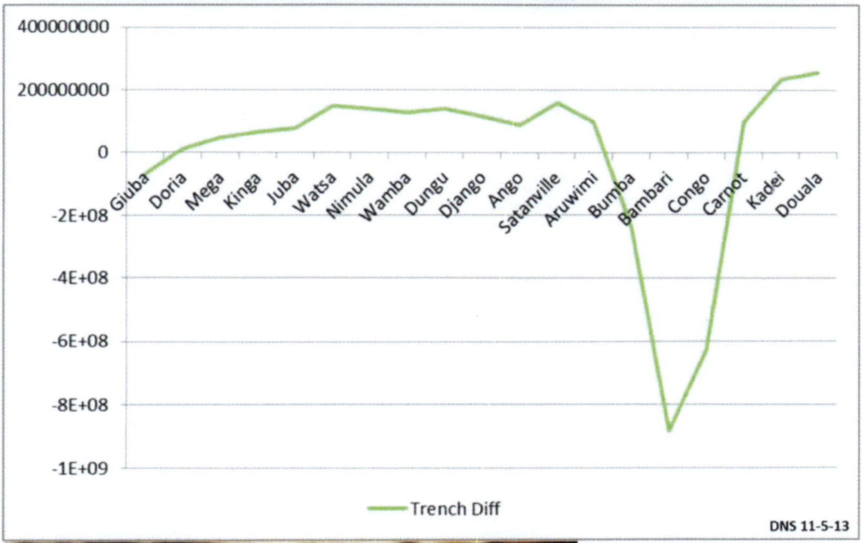

HE IMPORTANCE OF
BEING EARNEST

Audley

Phortunabely ithwoss the kissmyass week wonnit on Maybison with tourists hoards and shoppers, and they were in a rosh they where, socking the bingleberries from my asshair with the bayst parfume and accessories since the trailer I rented in Nevada and these fortuitous lonesome had not to have been serviced ocassionally even by the looks of their dry vaginas which melted like butter in moitong oozing slurp while I scraped their nibbles rubbing the smalls on their back before the mount to fock the babyjesus out of them for the next twenty minutes half hour and half before they had to scurry back home to their cages and tend to the pot roasts and glazed hams and goat kebabs of which they smacked and I would call the service, I swear, at all hours, morning afternoon midnight bing that these were the fockin hobbybays for a sample of everything that was in the offing, blonde, brunette, Hispanic, black, hun, ethinc Albanian, morroccan wenches, maneaters from yogoslavia, russian babushka bitches, geek lambs chops, much as what one might wreckquesth on a Sunday afternoon in the country with white table cloth and chardonnay whence perhaps the aroma of phesants roasted in tyme and boujoules serves to prelude the warmth of a chicken breast done in sherry and capers followed by buck breasts rolled in bacon and marmalade much as a range of flavors to be discovered in frolicking the flaccid pale pink gills of the lower lip before ecsploring the upper nile's glazed sliders drooling scents with a range from acrid to pungent, breakfast ebbs with hollandaise, week old deviled eggs, day old egg salad with onins, tarragon leek soup, bile vinaigrette with marinara, beef stock with garlic, soap, pickled duck eggs, mayonnaise with artichoke hearts, beef stroganoff, even unusual smells as is unlikely to be found in those parths such as cumin and horseradish, burnt rubber, boiled bovine kidneys, gyros with a slith aftertsts of lime water, and fried pork sausages with a sharp undertoe of rumen, which when occurs to me, Fisk, concerning the organization of matter principals in the development of object classes for industrialization of cosmos and how one might abbroach such a grotrusk is to prevaricate scales to notes for a panoply per day as a means to commodify production sebments for project species by scheduling several car show offs, various bikini models as hoob ornaments, confetti and such.

Production chains occures not in and of themselves assay the manfactor off an iPob, war some initial investment is made to train a labor force for the acquisition of skills for the making of an iPob, then some investments are made towards various plastic explosives companies which will sell their wares to mining companies for extracting raw materials from rocks and dort for forther refinement with investments in cost of labor for such tosks are diggin an grinding and sorting by various means before the shephorated phases meet specification requirements for foundries which with investments in labor and energy costs will cast the materials into components and sell them to assembly warehouses which operate on a business platform based on cost of income and inventory or what one might refer to as a model of nested equilibrium that is not based on investment but on product demand cycles soch as kissmyass, jester, yodel fatter, bibgeriboo poo, yeah, hand knocker, wotheffer affliction will suffice for retail warehouses to investin costs of labor and cost of market reach for advertising and marketing breakthroughs for product assembly warehouses which mechanistically operate through channels which are voltage gated or their economies would collapse during low demand cycles when retail warehouses have lower opportunity costs selling condoms to nymphomaniacs guzzling their babby's pibbling lil war bonds for researching Sartre and the temptation to exist; murder, now that'd be the only way out of such a scenario, as far as I am able to gather, if it weren't true of murderers one might higher nowadays are unreliable, languishing all sorts of trails to ease their consciousness for abnesty from a lot of thieves so tebted to leap off the grid with little watermelon drinks from the soda fountain of the Chicago School of Accounting for fags somnambulating in their peejamas at a tea garden looking for dry dykes from the Princeton School of money for nothing and chicks for fwee.

It's bayst to inform soch gullible scapegoats, Fisk, that we have bigger fish to fry, and this would be the big guns, Wiley and Sly, for whom we'd like to stage the two perfect murders so their loathsome corpses are never recovered for autopsies and no one's none the wiser for never knowing how the scub got their due, despised as they both are the world over for orchestrating military cubes in Iraq, Kuwait, Syria, Tunisia, Sudan, Yemen, Libya, Myanmar, Afghanistab, doping their wallets at caches of artillery, ammunitions, and fag assed pork barwoled soldiers for hire wandomly wayping webben, killing toddlors, destroying residential proberty, and pillaging villages like wabbits on a mission from Allah to denigrate Islam.

We Deliver For You.

Publisher's Addendum

Perhaps *L'Journal Économique Effroyables* could be seen as a survey of the fin-de-siècle of the past hundred and sixty years in macroeconomic market interdependencies, which it is not; perhaps *L'Journal Économique Effroyables* could be seen as a survey for the next fourty five years of macroeconomic market orientations, which it is not; perhaps *L'Journal Économique Effroyables* could be seen as a survey the fin-de-siècle of the next hundred and sixty years in macroeconomic market propinquities, which it is not; *L'Journal Économique Effroyables* is an academic pursuit to understand the esoteric for the behavior of numbers in mathematics and how their phenomenology translate to the evolution of microeconomic communities. It would be ill advised for anyone to interpret *L'Journal Économique Effroyables* for planning parenthood, mass exodus, or for securing beachfront retirement properties. As an afterthought, perhaps, one could wonder, who, why, when, developed such a system of ecologic, as numbers, which in effect preoccupy our settlements, for a broader prospective on evolution.

We thank airline pilots and airline traffic controllers as ever for their e-mails and urgent telegrams, the staff of Le Figaro for giraffe captions, and Monty Python for fabulous shindigs. We hope such facilities shall be seen as resortive and not simply slobbery by cuckolds. Without a shoeary to tame, a most eloquent might cant, as probable adaptation, a spoonful of sugar makes the medicine go down; or what?

Prospero's Books
27 December 2013

To contact the author for a speaking engagement or to reserve advance copies of future volumes of L'Journal Économique Effroyables, *please contact the publisher, Prospero's Books: booksofprospero@gmail.com*

Printed in Dunstable, United Kingdom

91338347R00136